PLANT BREEDING

PLANT BREEDING

By

D. R. Bhardawaj

2016

SBS Publishers & Distributors Pvt. Ltd.

New Delhi

ISBN 13 : 9789380090672

First Published in 2016

Published by:

SBS PUBLISHERS & DISTRIBUTORS PVT. LTD.

2/9, Ground Floor, Ansari Road, Darya Ganj,

New Delhi - 110002,

INDIA

Tel: 0091.11.23289119 / 41563911

Email: mail@sbspublishers.com

www.sbspublishers.com

Preface

Plant breeding is the art and science of changing the traits of plants in order to produce desired characteristics. It can be accomplished through many different techniques ranging from simply selecting plants with desirable characteristics for propagation, to more complex molecular techniques. Plant breeding has been practiced for thousands of years, since near the beginning of human civilization. It is now practiced worldwide by individuals such as gardeners and farmers, or by professional plant breeders employed by organizations. Plant breeder's right is also a major and controversial issue. Today, production of new varieties is dominated by commercial plant breeders, who seek to protect their work and collect royalties through national and international agreements based in intellectual property rights. The range of related issues is complex. In the simplest terms, critics of the increasingly restrictive regulations argue that, through a combination of technical and economic pressures, commercial breeders are reducing biodiversity and significantly constraining individuals from developing and trading seed on a regional level. Efforts to strengthen breeders' rights, for example, by lengthening periods of variety protection are ongoing. Classical plant breeding uses deliberate interbreeding of closely or distantly related individuals to produce new crop varieties or lines with desirable properties. Modern plant breeding uses techniques of molecular biology to select, or in the case of genetic modification, to insert, desirable traits into plants.

Editor

Contents

Chapter 1

UNLOCKING THE GENETIC DIVERSITY OF MAIZE LANDRACES WITH DOUBLED HAPLOIDS OPENS NEW AVENUES FOR BREEDING

Alexander Strigens[1], Wolfgang Schipprack[1], Jochen C. Reif[2], Albrecht E. Melchinger[1*]

[1]Institute of Plant Breeding, Seed Science and Population Genetics, University of Hohenheim, Stuttgart, Germany

[2]State Plant Breeding Institute, University of Hohenheim, Stuttgart, Germany

ABSTRACT

Landraces are valuable genetic resources for broadening the genetic base of elite germplasm in maize. Extensive exploitation of landraces has been hampered by their genetic heterogeneity and heavy genetic load. These limitations may be overcome by the in-vivo doubled haploid (DH) technique. A set of 132 DH lines derived from three European landraces and 106 elite flint (EF) lines were genotyped for 56,110 single nucleotide polymorphism (SNP) markers and evaluated in field trials at five locations in Germany in 2010 for several agronomic traits. In addition, the landraces were compared with synthetic populations produced by intermating DH lines derived from the respective landrace. Our objectives were to (1) evaluate the phenotypic and molecular diversity captured within DH lines derived from European landraces, (2) assess the breeding potential (usefulness) of DH lines derived from landraces to broaden the genetic base of the EF germplasm, and (3) compare the performance of each landrace with the synthetic population produced from the respective DH lines. Large genotypic variances among DH lines derived from landraces allowed the identification of DH lines with grain yields comparable to those of EF lines. Selected DH lines may thus be introgressed into elite germplasm without impairing its yield level. Large genetic distances of the DH lines to the EF lines demonstrated the potential of DH lines derived from landraces to broaden the genetic base of the EF germplasm. The comparison of landraces with their respective synthetic population showed no yield improvement and no reduction of phenotypic diversity. Owing to the low population structure and rapid decrease of linkage disequilibrium within populations of DH lines derived from landraces, these would be an ideal tool for association mapping. Altogether, the DH technology opens new opportunities for characterizing and utilizing the genetic diversity present in gene bank accessions of maize.

INTRODUCTION

According to molecular evidence, maize (Zea mays L.) was introduced into Europe over two distinct paths. A first introduction of Caribbean Flint maize in Spain by Colombus in 1493 was followed by introductions of Northern Flints from North America to North-Western Europe during the 16th century [1]. This flint maize was cultivated as open-pollinated populations in different regions of Europe over centuries, resulting in a broad diversity of landraces. Natural selection promoted adaptation to the cool and wet climatic conditions prevailing in large parts of the continent [2], while artificial selection through saving desirable ears for the next growing season shaped the landraces according to farmers' preferences. Hybridization of flint populations from Spain with those from the northern introduction in the Pyrenean region produced a novel germplasm that was most likely the cradle of the European elite flint (EF) germplasm [3]. Despite good adaptation of landraces to local climatic and soil conditions, the advent of highly productive hybrids in the 1960s resulted in a rapid decline of their cultivation [4]. Fortunately, the value of landraces as genetic resources was recognized before their extinction. They were collected at their growing locations and are being conserved ex situ, in gene banks.

During the early cycles of hybrid breeding in the 1950's and 1960's, a portion of the genetic diversity and specific adaptation of the European landraces was captured in the founder lines of the modern European flint heterotic pool [5]. However, only few inbred lines derived from a few landraces contributed to the elite breeding material. Broadening the genetic base of the elite breeding pool through introgression of additional material from European landraces might, therefore, be of great interest [6]. Extensive characterization studies of landraces for morphological traits [10], [11], early growth and cold tolerance [12]–[16], adaptation to low nitrogen fertilization [17], [18], and pest resistance [19], [20], often identified landraces with desirable traits. Germplasm collections of European landraces were also extensively genotyped with molecular markers to determine their origin, relatedness, and degree of genetic diversity [1], [7]–[9], which revealed their broad genetic diversity and potential for further mining of favorable alleles. Nevertheless, the presence of deleterious and recessive lethal alleles in the landraces, commonly referred to as genetic load or genetic burden, has so far hampered their direct use in breeding [21].

Inbreeding uncovers the genetic load and can be used to purge landraces from lethal or detrimental alleles [22]. However, line development by recurrent selfing is extremely cumbersome. Recessive lethal alleles masked in heterozygous plants during early generations result in losses of lines in advanced selfing generations, due to increasing homozygosity. Use of the in-vivo doubled haploid (DH) technique [23] was proposed to overcome these drawbacks [7]. This technique is routinely used for line development in commercial breeding programs in Europe and the US and could offer several advantages with respect to the exploitation of landraces. First, recessive lethal alleles are expressed already at the haploid stage, which should reduce a part of the genetic load at the very beginning of line development [24], [25]. Second, alleles present in heterogeneous populations are fixed in a single step in homozygous lines, ideally capturing most of the variation present in the original source [26]. Third, DH lines can be reproduced ad libitum and evaluated with any desired degree of precision in replicated trials, whereas landraces in any generation represent a conglomerate of highly diverse, unique and non-reproducible individuals. This enables maintenance and efficient selection of the most promising genotypes for further use in breeding.

First studies using the in-vivo DH technique to produce fixed lines from European and tropical landraces showed a generally lower testcross performances but also a high improvement potential of such materials [18], [27]. Yet, information about the per se performance of a broad set of DH lines derived from landraces would be of equal interest with regard to their introgression into modern breeding material. In addition to an evaluation of the phenotypic diversity captured in these lines, it would allow selection of agronomic traits relevant for line maintenance and hybrid seed production [28]. Evaluation of the molecular diversity among these DH lines and its comparison with that of the EF material could provide important information about the level of untapped genetic variation present in the landraces.

The goals of this study were to (1) evaluate the phenotypic and molecular diversity captured within DH lines derived from European landraces and compare them with the diversity of EF lines, (2) assess the breeding potential (usefulness) of DH lines derived from landraces to broaden the genetic base of the EF germplasm, and (3) compare the performance of each landrace with the synthetic population produced from their respective DH lines. The comparison of DH lines derived from the landraces Bugard, Gelber Badischer and Schindelmeiser with EF lines in field trials and marker assays revealed substantial phenotypic and molecular variation within the populations of DH lines, supporting their potential to broaden the genetic base of the EF germplasm.

RESULTS

Experiment 1 (DH lines from landraces and elite material)

Mean performance reveals diversity among populations.

Significant differences ($P<0.05$) were observed among the means of the populations of DH lines from landraces and the EF lines for all morphological and agronomic traits except female flowering (Table 1), as well as for all yield components (Table 2). The DH lines derived from the landrace Bugard (DH-BU) had on average a low emergence rate, high leaf chlorosis, low early growth rate and short anthesis-silking interval. They had short plants with intermediate ear height, well developed ear shanks and almost no husk flag leaves. The plants carried short broad ears with 10.8 kernel rows and 16.4 kernels per row on average, with a tendency to develop a second seed setting ear.

The DH lines derived from the landrace Gelber Badischer (DH-GB) had a high emergence rate, low leaf chlorosis, high early growth rates and an intermediate anthesis-silking interval. The tall plants with high ear insertion also had long ear shanks and husk flag leaves. Their ears were long and slender, with 8.4 kernel rows and 20.3 kernels per row. The DH-GB lines showed the highest mean for 100 kernel weight among all populations. The DH lines derived from the landrace Schindelmeiser (DH-SC) showed values intermediate to those of DH-BU and DH-GB for most of the traits. Exceptions were ear height, ear shank and ear dry matter content, where DH-SC lines had lower means than DH-BU and DH-GB. In contrast, the DH-SC lines showed higher means for anthesis-silking interval and kernel rows.

The EF lines had on average 22% higher grain yield (55.2 g plant−1) than the DH lines derived from landraces (43.0 g plant−1). Grain yield and number of kernel rows were the only traits for which the EF lines significantly ($P<0.05$) outperformed all three populations of DH lines. The founder lines EP1, F2, F7 and DK105 had a grain yield of 49.5, 38.9, 47.2 and 55.0 g plant−1, respectively, and were thus below the average grain yield of the EF lines.

Table 1. Mean, genotypic variance (σ2g), genotype×environment interaction variance (σ2ge), heritability (h2), predicted gain from selection (ΔG) and usefulness (U) at selection intensity α for agronomic and morphological traits of doubled haploid (DH) lines derived from the landraces Bugard (DH-BU, n = 36), Gelber Badischer (DH-GB, n = 31) and Schindelmeiser (DH-SC, n = 65), as well as of elite flint lines (EF, n = 106).

						α = 10%		α = 40%	
Trait	**Population**	**Mean**†	σ^2_g	σ^2_{ge}	h^2	ΔG	U	ΔG	U
Emergence %	DH-BU	52.4a	217.72**	47.44**	0.94	25.18	77.58	13.88	66.28
	DH-GB	63.3bc	67.59**	22.95**	0.88	13.57	76.87	7.48	70.78
	DH-SC	58.8ab	134.17**	26.29**	0.93	19.66	78.46	10.84	69.64
	EF	64.8b	113.47**	42.36**	0.89	17.69	82.49	9.75	74.55
Leaf chlorosis 1–9‡	DH-BU	4.20b	0.81**	1.02**	0.76	1.38	2.82	0.76	3.44
	DH-GB	3.15a	0.46**	0.35**	0.79	1.06	2.09	0.58	2.57
	DH-SC	3.19a	0.40**	0.42**	0.74	0.96	2.23	0.53	2.66
	EF	3.10a	0.33**	0.28**	0.74	0.87	2.23	0.48	2.62
Relative growth rate GDD $^{-1}$×10^{-3}§	DH-BU	17.2a	0.61*	0.99**	0.66	1.12	18.32	0.62	17.82
	DH-GB	18.9b	0.66**	0.00	0.84	1.31	20.21	0.72	19.62
	DH-SC	18.3b	0.91**	0.21*	0.85	1.55	19.85	0.85	19.15
	EF	17.6a	0.56**	0.68**	0.69	1.09	18.69	0.60	18.20
Female flowering GDD	DH-BU	661a	1947**	374**	0.94	75	736	41	702
	DH-GB	659a	1806**	246**	0.95	73	732	40	699
	DH-SC	656a	1386**	398**	0.92	63	719	35	691
	EF	638a	1902**	239**	0.96	75	713	41	679
Anthesis-silking interval GDD	DH-BU	31.8a	209**	126**	0.81	23	8.8	13	18.8
	DH-GB	42.3ab	430**	163**	0.88	34	8.3	19	23.3
	DH-SC	52.9b	404**	188**	0.86	33	19.9	18	34.9
	EF	34.1a	388**	83**	0.91	33	1.1	18	16.1
Plant height cm	DH-BU	134a	355**	40**	0.96	32	166	18	152
	DH-GB	164c	432**	50**	0.96	36	200	20	184
	DH-SC	144ab	329**	31**	0.96	31	175	17	161
	EF	149b	313**	16**	0.97	31	180	17	166
Ear height cm	DH-BU	49.2ab	144.94**	9.90*	0.96	20.76	69.96	11.44	60.64
	DH-GB	53.3b	63.37**	23.24**	0.86	12.99	66.29	7.16	60.46
	DH-SC	44.1a	73.95**	9.64**	0.92	14.52	58.62	8.00	52.10
	EF	51.5b	76.26**	8.10**	0.93	14.82	66.32	8.17	59.67
Ear shank 1–9‡	DH-BU	4.48b	1.38**	0.26**	0.92	1.98	2.50	1.09	3.39
	DH-GB	4.45b	1.84**	0.08	0.96	2.34	2.11	1.29	3.16
	DH-SC	3.75a	0.92**	0.38**	0.86	1.57	2.18	0.86	2.89
	EF	4.08ab	0.77**	0.06**	0.92	1.48	2.60	0.82	3.26
Husk flag leaves 1–9‡	DH-BU	1.74a	1.57**	0.00	0.96	2.16	−0.42	1.19	0.55
	DH-GB	4.19c	1.75**	1.16**	0.83	2.12	2.07	1.17	3.02
	DH-SC	3.50b	2.04**	0.94**	0.87	2.34	1.16	1.29	2.21
	EF	1.60a	0.44**	0.01	0.88	1.10	0.50	0.60	1.00

*, **Significant at the 0.05, 0.01 probability level, respectively.

†Different letters indicate significant differences among the four populations for the respective trait.

‡1 = absent, 9 = pronounced.

§Multiply the reported mean by this this value and the variance components by the square of this this value to obtain the actual numbers.

doi:10.1371/journal.pone.0057234.t001

doi:10.1371/journal.pone.0057234.t001

Table 2. Mean, genotypic variance (σ2g), genotype×environment interaction variance (σ2ge), heritability (h2), predicted gain from selection (ΔG) and usefulness (U) at selection intensity α for grain yield and yield components of doubled haploid (DH) lines derived from the landraces Bugard (DH-BU, n = 36), Gelber Badischer (DH-GB, n = 31) and Schindelmeiser (DH-SC, n = 65), as well as of elite flint lines (EF, n = 106).

						α = 10%		α = 40%	
Traits	Population	Mean[†]	σ^2_g	σ^2_{ge}	h^2	ΔG	U	ΔG	U
Ear length cm	DH-BU	10.7a	1.54**	0.14	0.90	2.07	12.77	1.14	11.84
	DH-GB	14.1c	2.88**	1.21**	0.87	2.79	16.89	1.54	15.64
	DH-SC	12.7b	3.39**	1.05**	0.9	3.07	15.77	1.69	14.39
	EF	13.3bc	2.39**	0.29**	0.92	2.61	15.91	1.44	14.74
Ear diameter mm	DH-BU	34.3b	8.30**	1.47**	0.92	4.86	39.16	2.68	36.98
	DH-GB	30.7a	5.98**	0.48	0.92	4.13	34.83	2.28	32.98
	DH-SC	34.5b	8.75**	1.81**	0.91	4.97	39.47	2.74	37.24
	EF	33.9b	4.33**	0.25	0.91	3.49	37.39	1.93	35.83
Kernel rows #	DH-BU	10.8b	1.41**	0.11*	0.94	2.03	12.83	1.12	11.92
	DH-GB	8.4a	0.35**	0.00	0.85	0.96	9.36	0.53	8.93
	DH-SC	11.6c	1.18**	0.19**	0.91	1.82	13.42	1.01	12.61
	EF	12.9d	1.29**	0.14**	0.93	1.93	14.83	1.06	13.96
Kernels per row #	DH-BU	16.4a	6.71**	0.70	0.86	4.23	20.63	2.33	18.73
	DH-GB	20.3b	13.63**	7.20**	0.84	5.96	26.20	3.28	23.58
	DH-SC	16.4a	12.56**	3.63**	0.88	5.85	22.25	3.22	19.62
	EF	21.6b	8.07**	1.44**	0.87	4.66	26.26	2.57	24.17
Hundred kernel weight g	DH-BU	22.1a	14.20**	4.83**	0.90	6.29	28.39	3.47	25.57
	DH-GB	25.0b	6.05**	4.61**	0.79	3.85	28.85	2.12	27.12
	DH-SC	22.9a	11.46**	2.54**	0.91	5.68	28.58	3.13	26.03
	EF	21.1a	7.73**	0.72**	0.93	4.72	25.82	2.60	23.7
Ear dry matter content[‡] %	DH-BU	57.4bc	20.07**	5.32**	0.93	7.60	65.00	4.19	61.59
	DH-GB	56.1ab	10.40**	5.30**	0.86	5.26	61.36	2.90	59.00
	DH-SC	53.0a	46.97**	6.49**	0.96	11.82	64.82	6.51	59.51
	EF	59.1c	11.12**	2.00**	0.94	5.69	64.79	3.14	62.24
Grain yield g plant^{-1}	DH-BU	42.5a	82.68**	19.55*	0.84	14.67	57.17	8.08	50.58
	DH-GB	44.9a	65.22**	27.32*	0.79	12.63	57.53	6.96	51.86
	DH-SC	41.7a	114.47**	19.91**	0.88	17.66	59.36	9.74	51.44
	EF	55.2b	82.42**	21.61**	0.83	14.56	69.76	8.02	63.22

*, **Significant at the 0.05, 0.01 probability level, respectively.
[†]Different letters indicate significant differences among the four populations for the respective trait.
[‡]at 420 GDD after flowering.
doi:10.1371/journal.pone.0057234.t002

doi:10.1371/journal.pone.0057234.t002

Grain yield was significantly (P<0.05) associated with all yield components (Table 3), with the exception of 100 kernel weight for DH-GB. The number of kernels per row explained on average 42% of the phenotypic variation in grain yield, while ear length accounted for 23% only. Grain yield was positively associated with early growth rates of the DH-GB lines, and with plant height of the EF lines, while negative associations were observed with female flowering of the

DH-SC lines and with anthesis-silking interval of the DH-GB lines. Ear dry matter content was positively associated with grain yield of the DH-GB lines, whereas the association was negative for the EF lines.

Table 3. Correlations of agronomic and morphologic traits with grain yield per plant within populations of doubled haploid (DH) lines derived from the landraces Bugard (DH-BU, n = 36), Gelber Badischer (DH-GB, n = 31) and Schindelmeiser (DH-SC, n = 65) as well as within elite flint lines (EF, n = 106).

Trait	DH-BU	DH-GB	DH-SC	EF
Emergence	0.31	0.36	−0.21	−0.20
Leaf chlorosis	−0.22	−0.23	−0.04	−0.03
Relative growth rate	0.27	0.51**	0.16	0.10
Female flowering	−0.15	−0.22	−0.33**	−0.03
Anthesis-silking interval	0.24	−0.39*	−0.17	−0.17
Plant height	0.13	0.09	0.23	0.41**
Ear height	0.33	0.07	0.1	0.18
Ear shank	0.03	−0.02	0.13	0.04
Husk flag leaves	−0.27	0.03	0.04	−0.07
Ear length	0.42*	0.41*	0.64**	0.41**
Ear diameter	0.61**	0.55**	0.58**	0.70**
Kernel rows	0.33*	0.42*	0.25*	0.45**
Kernels per row	0.69**	0.73**	0.71**	0.42**
Hundred kernel weight	0.45**	0.24	0.45**	0.42**
Ear dry matter content	−0.05	0.42*	0.23	−0.28**

*, **Significant at the 0.05, 0.01 probability level, respectively.

doi:10.1371/journal.pone.0057234.t003

doi:10.1371/journal.pone.0057234.t003

The average pair-wise phenotypic distance was 5.30 among the DH-BU lines, 4.93 among the DH-GB lines, 5.22 among the DH-SC lines, and 4.44 among the EF lines (Figure 1). Average phenotypic distance of the EF lines to the populations of DH lines was 5.72 for DH-BU, 5.97 for DH-GB, and 5.58 for DH-SC.

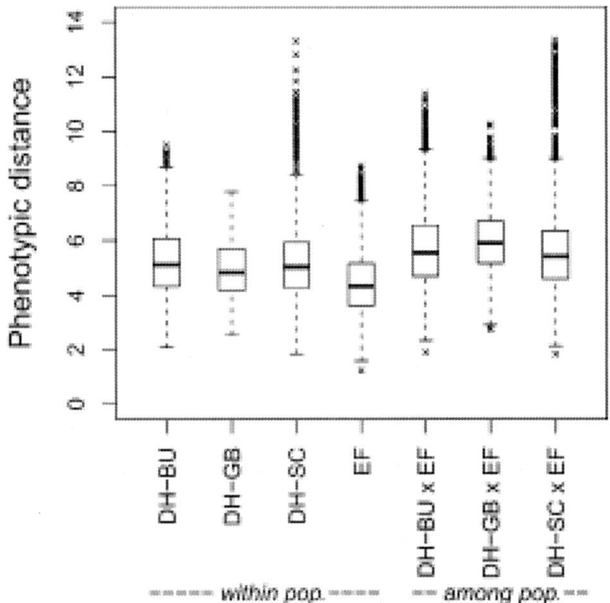

Figure 1. Phenotypic distances among elite flint lines and doubled haploid lines derived from landraces.

Pair-wise Euclidean distances among elite flint (EF) lines and doubled haploid (DH) lines derived from the landraces Bugard (DH-BU), Gelber Badischer (DH-GB), and Schindelmeiser (DH-SC) were calculated from 16 morphological and agronomic traits standardized to mean zero and unit variance.

doi:10.1371/journal.pone.0057234.g001

Large genotypic variances within all populations.

Estimates of genotypic variance were significant ($P<0.05$) for all populations and traits (Tables 1 and 2). In most instances, the genotypic variance was higher for the populations of DH lines derived from landraces than for the EF lines. For the DH-BU lines, markedly low estimates of genotypic variance were observed for anthesis-silking interval, ear length and number of kernels per row. For the DH-GB lines, the estimates of genotypic variance were low for emergence rate, kernel rows, 100 kernel weight, ear dry matter content, and grain yield. The EF lines showed the lowest genotypic variance for leaf chlorosis, early growth rate, plant height, ear diameter, ear shank and husk flag leaves scores. Despite generally high estimates of genotype×environment interaction variance, heritabilities were moderate for growth rates of the DH-BU and EF lines (0.66 and 0.69, respectively), and high to very high (0.74 to 0.97) for the remaining traits. In most instances, the usefulness criterion at a selected fraction of $\alpha = 40\%$ (U40%) of the populations of DH lines derived from landraces surpassed the mean of the EF lines. Important exceptions were grain yield for all populations of the DH lines and the number of kernel rows for DH-GB lines, as none of them had more than 10 kernel rows. The usefulness criterion at a selected fraction of $\alpha = 10\%$ (U10%) of the populations of DH lines reached in many instances the usefulness of the EF lines. The top 10% DH lines (DH10%) selected based on index performance across all three DH line populations had an average grain yield of 54.05 g plant−1 while the top 10% EF lines (EF10%) selected according to the same index had an average grain yield of 69.43 g plant−1.

Molecular analyses reveal high genetic variation within populations.

Mean pair-wise modified Rogers' distance (dW) between lines calculated from 24,572 SNP markers was highest in DH-BU (0.52), and lowest in DH-SC (0.45) (Figure 2). The range of pair-wise distances was similar within the three populations of DH lines with dW values from 0.21 to 0.56. By comparison, the mean dW between the EF lines was 0.50, with values ranging from 0.03 to 0.66. Average dW between the EF lines and populations of DH lines was highest for DH-BU (0.61) and lowest for DH-GB (0.58). The minimum dW between elite lines and DH lines derived from landraces was observed between the EF line DK105 and a DH-GB line (0.48).

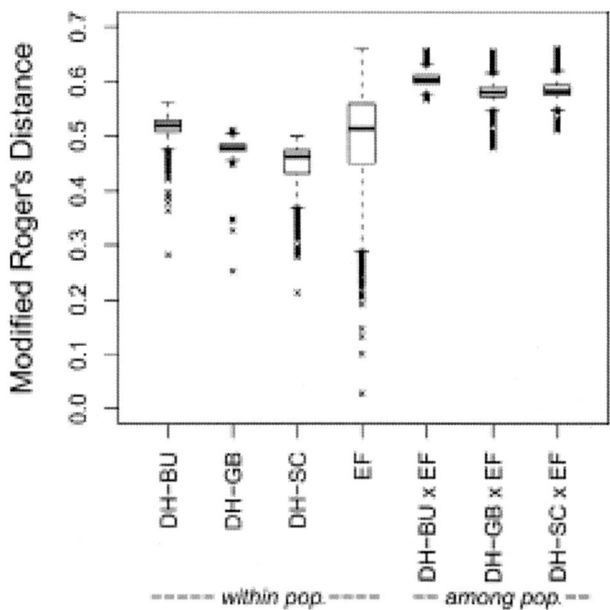

Figure 2. Genetic distances among elite flint lines and double haploid lines derived from landraces.

Pair-wise modified Rogers' distances among elite flint (EF) lines and doubled haploid (DH) lines derived from the landraces Bugard (DH-BU), Gelber Badischer (DH-GB), and Schindelmeiser (DH-SC), based on 24,572 SNP markers.

doi:10.1371/journal.pone.0057234.g002

The number of unique alleles was highest in the EF. In total, 1,556 SNP markers were absent from the three populations of DH lines derived from landraces (Table 4), thus representing on average 16.04 unique alleles per an EF line. Among the populations of DH lines, the average number of unique alleles per line was highest (7.64) in DH-BU and lowest (0.95) in DH-SC. Nei's genetic diversity within population (Hwi) was similar for the EF and DH-BU materials, whereas DH-GB and DH-SC showed lower values (Table 4). Average within population genetic diversity (Hw) was 0.233 and the total genetic diversity among all flint lines (Ht) was 0.312, resulting in a mean genetic differentiation (GST) of 0.253. Linkage disequilibrium (LD) decreased below the threshold of r2 = 0.1 within less than 0.8

Mbp in the populations of DH lines derived from landraces but only after 3.8 Mbp within the EF lines (Table 4).

Table 4. Number of lines (N) and population specific SNPs, genetic diversity within population (Hwi), population differentiation (GSTi) and extent of linkage disequilibrium (LD) within populations of doubled haploid (DH) lines derived from the landraces Bugard (DH-BU), Gelber Badischer (DH-GB) and Schindelmeiser (DH-SC) as well as within elite flint lines (EF).

	DH-BU	DH-GB	DH-SC	EF
N	36	29	60	97
Specific SNPs	275	87	57	1556
H_{wi}	0.260	0.221	0.200	0.253
G_{STi}	0.169	0.292	0.359	0.189
LD Mbp	0.725	0.275	0.375	3.875

doi:10.1371/journal.pone.0057234.t004

doi:10.1371/journal.pone.0057234.t004

Experiment 2 (landraces and synthetic populations)

The landraces Bugard, Gelber Badischer and Schindelmeiser differed significantly in their mean performance ($P<0.05$) for most of the traits, with the exception of emergence rate, leaf chlorosis score, kernels per row and 100 kernel weight (Table 5). Bugard had the lowest early growth rate and early female flowering, and showed a narrow anthesis-silking interval. The plants were of small stature with a high ear insertion, and carried short broad ears with an average of 11.1 kernel rows and without any husk flag leaves. The high grain yield obtained with Bugard mainly resulted from a high number of kernels per plant. Gelber Badischer had a higher early growth rate, later female flowering and a wider anthesis-silking interval. The plants were tall, with similar ear height as Bugard and carried long slender ears with 8.7 kernel rows on average. They showed low grain yield and the lowest ear dry matter content. Schindelmeiser was intermediate to Bugard and Gelber Badischer for most traits. Only ear height was markedly lower than for the other two landraces.

Table 5. Mean performance for agronomic and morphological traits of the landraces Bugard, Gelber Badischer and Schindelmeiser, as well as of the corresponding synthetic populations produced by intermating the respective doubled haploid lines.

	Bugard		Gelber Badischer		Schindelmeiser		
Trait	Landrace	Synthetic	Landrace	Synthetic	Landrace	Synthetic	Synthetic effect
Emergence %	61.8a[†]	62.2a	62.2a	65.3a	62.0a	62.3a	
Leaf chlorosis 1–9[‡]	2.81ab	3.75b	2.59ab	2.53ab	2.27a	2.80ab	
Relative growth rate $GDD^{-1} \times 10^{-3}$[§]	18.5ab	18.1a	20.0c	19.8c	19.3bc	18.9ab	
Female flowering GDD	605a	605a	675c	640b	631b	642b	**
Anthesis-silking interval GDD	44.7a	42.0a	89.5c	69.1b	68.6b	75.2b	*
Plant height cm	152a	161b	191d	176c	157ab	161b	
Ear height cm	69.8b	71.3b	74.5c	67.5b	56.0a	53.6a	**
Ear shank 1–9[‡]	4.0a	4.5b	4.6b	4.6b	3.8a	3.5a	
Husk flag leaves 1–9[‡]	1.0a	1.1a	1.9b	2.3c	1.5b	1.7b	
Ear length cm	13.5a	13.8a	17.2bc	18.9c	16.2b	15.5ab	
Ear diameter mm	39.5c	38.6c	34.7b	31.8a	35.7b	36.0b	*
Kernel rows #	11.1c	10.3b	8.7a	8.2a	12.1c	12.0c	*
Kernels per row #	26.8ab	28.0ab	30.1ab	31.8b	26.0a	26.0a	
Hundred kernel weight g	27.9b	27.0b	26.9b	26.9b	25.9ab	24.5a	
Kernel dry matter content %	66.0b	65.5b	61.7a	65.2b	65.1b	65.4b	**
Grain yield g $plant^{-1}$	64.3b	68.2b	47.1a	48.4a	51.1a	50.3a	
Kernels per plant	227cd	252d	170a	178ab	195abc	215bc	*
Barren stalks %	2.41ab	0.00a	13.83c	7.63bc	7.47b	2.78ab	**
Common smut %	11.61a	8.41a	21.05b	13.61ab	7.34a	5.87a	*

[†]Different letters indicate significant differences among populations for the respective trait.
[‡]1 = absent, 9 = pronounced.
[§]Multiply the reported mean by this value to obtain the actual numbers.
doi:10.1371/journal.pone.0057234.t005

doi:10.1371/journal.pone.0057234.t005

In the comparison of landraces with their corresponding synthetic population, significant (P<0.05) and systematic differences were observed for female flowering, anthesis-silking interval, ear height, ear diameter, number of kernel rows, kernel dry matter content, number of kernels per plant, number of barren stalks and incidence of common smut (Table 5). No improvement of grain yield was observed in the synthetic populations compared to the source landraces. Individual synthetic populations differed significantly (P<0.05) from their respective source landraces for plant height, ear shank and husk flag leaves score. Estimates of residual variance among individuals decreased significantly (P<0.05) in the synthetic population produced from the DH-BU lines compared to Bugard for plant height. Similarly, the same trend was observed in the synthetic population produced from the DH-GB lines compared to Gelber Badischer for number of kernel rows. Increased estimates of residual variance among individuals were observed for husk flag leaves in all synthetics (data not shown). No significant differences between estimates of residual variance among individuals were observed for the remaining traits.

DISCUSSION

Recovering the diversity of landraces for breeding

Unlocking the genetic diversity present in the huge collections of maize landraces accessions by developing homozygous lines from these highly heterogeneous populations [18], [27] is of great interest for breeding. Both the single seed descent (SSD) and the *in vivo* double haploid (DH) techniques would be appropriate methods to produce fixed lines and purge the genetic load present in landraces. According to a roughly ten-fold lower efficiency of DH production when landraces are used as source population in comparison to elite crosses (Schipprack, personal communication), and following the reports on the development of first cycle inbred lines [29], high losses (~90%) of lines due to the genetic load can be expected during the selfing or DH production process. Thus, to produce a sufficient number of unrelated inbred lines, huge numbers of individuals are needed from each population to start line development. While it is rather easy to pollinate large numbers of plants with inducers in isolation fields for a large scale DH production, producing 150 inbred lines by SSD would require selfing of around 1,500 individual plants for at least six generations. In addition, since heterozygous plants are more vigorous and as such preferred for SSD line derivation, it might even require up to eight selfing generations to produce lines with an acceptable level of homozygosity [30]. Given that only one kernel per ear is used for the next generation, approximately 12,000 hand pollinations would be required for the whole process. In comparison, assuming that 20% of the viable haploid plants produce seeds [23], only ~3,000 hand pollinations would be required for line development (1 ear) and multiplication (3 ears) when using the DH technique. Even though chromosome doubling and selfing require additional efforts, this still results in considerably less work. Owing to the higher expected costs of SSD line development, we only used the DH technique to derive lines from the landraces, and could not compare these two methods with regard to their effect on the genetic diversity recovered.

Legitimate concerns on the selective neutrality of the *in vivo* DH technique could be raised because of the segregation distortion observed in DH lines produced by *in vitro* anther culture[31]. However, first studies on DH lines derived from single crosses of elite lines by the *in vivo*maternal DH production technique showed no systematic effect on phenotypic and allelic distribution [32], [33]. We could not monitor potential changes in the allele frequencies caused by the DH technique because the landraces themselves were not genotyped. However, to evaluate potential losses in genetic diversity, we compared the landraces with synthetic populations produced by intermating DH lines derived from the respective landrace in field trials. No bottleneck could be observed, as the residual variance among individuals was generally similar in the synthetic populations and landraces despite significant differences between mean performances for some traits (Table 5). Small but inconsistent differences in the means between the synthetic populations and the landraces (*e.g.*, plant height, husk flag leaves) might reflect random drift due to the small number of DH lines (DH-BU and DH-GB) extracted from landraces. In contrast, systematic effects (*e.g.*, female flowering, kernels per plant, number of barren stalks) suggested loss of specific alleles during the DH production or by selection during line multiplication.

Although improved seed set, reduced number of barren stalks, reduced anthesis-silking interval and lower incidence of common smut might suggest some positive purging of detrimental alleles[34], [35], the absence of grain yield improvement in the synthetic populations was disappointing and contrary to expectations [25], [27]. It must be noted, however, that seeds of the synthetic populations were produced on less vigorous homozygous lines while the seeds of the landraces were produced on vigorous heterozygous plants. It remains unclear whether purging of recessive alleles had no effect on grain yield of the resynthesized populations or whether this comparison was confounded by maternal effects [36]. Intermating each synthetic population for one generation and a comparison between the Syn-2 generation and the corresponding landrace would be necessary to resolve this issue.

Large diversity in DH lines from landraces in comparison with EF lines As indicated by the estimates of genotypic variances (Tables 1 and 2) and H_{wi} (Table 4), both the phenotypic and molecular diversity within the three populations of DH lines derived from landraces were comparable to that of the EF lines. Having in mind both a high G_{ST} of the DH-GB and DH-SC, and the fact that the EF germplasm is a mixture of various European and Northern-American materials, we can conclude that the diversity recovered from individual landraces was substantial. The H_{wi} values from the populations of DH lines were even similar to those obtained for much larger commercial breeding populations [37]. Considering that estimates of H_{wi} obtained from SNP marker data are approximately half than those obtained with simple sequence repeat (SSR) markers [37], [38], the H_{wi} of the populations of DH lines were comparable to values reported for European landraces in previous studies [7], [8]. This suggests no noticeable loss of molecular diversity during the production of DH lines.

Most striking was the narrow range of d_w values within the populations of DH lines compared to the one observed within the EF material (Figure 2). Except for a few pairs of lines showing d_w values at half of the population mean, most likely because they originated from the same female plant in the induction crosses, all DH lines were nearly equally related. This underlines the very low population structure within the landraces in comparison with that of the EF material, suggesting that the effective population size N_e of the landraces was much higher than the number of lines extracted, as re-sampling of gametes was negligible. Moreover, it supports a random sampling of gametes during DH line extraction, and underlines the value of the DH method to extract large numbers of unrelated lines from landraces.

The strongest phenotypic difference between the EF lines and the populations of DH lines derived from landraces was observed for grain yield (Table 2). With on average 22% less grain yield, the gap between the unselected DH lines and the EF lines was substantial and similar to values obtained when testcrosses of DH lines from *Gelber Badischer* and *Schindelmeiser* (22 to 26% less) were compared with commercial hybrids [18]. Grain yield of the DH lines was more tightly associated with the number of kernels per row than with ear length (Table 3). Comparison between the DH-SC and EF with similar ear size pointed to poor seed set as the main cause of the reduced grain yield of the DH lines. Trends for reduced anthesis-silking intervals, shorter husk flag leaves as well as larger ears with better seed set in modern vs. old lines (Table 1 and2) were also reported for parental inbred lines of U.S. hybrids released between 1930 and 2000[39]. This presumably reflects a correlated response of modern breeding germplasm to selection for grain yield under higher planting densities and led to the conclusion that the selection of lines with superior *per se* performance under stress conditions will also result in higher-yielding hybrids [40], [41].

Potential of DH lines from landraces for broadening the genetic base of the elite flint pool

The usefulness criterion $U(a)$ showed that the generally lower performance level of the populations of DH lines was largely compensated by large estimates of genotypic variance (Tables 1 and 2). Thus, detrimental agronomic properties (*e.g.*, lodging, poor seed set) present in the landraces can be removed prior to introgression into EF matrerials by selecting superior lines among the DH lines derived from landraces. Moreover, the yield gap can be substantially reduced because the top $DH_{10\%}$ lines reached nearly the mean grain yield performance of the EF lines. Similarly, a selection of the best performing DH lines reduced the yield gap between commercial hybrids and testcrosses of DH lines derived from *Gelber Badischer* and*Schindelmeiser* by 50% [18]. However, crosses of $DH_{10\%}$ lines with EF will still result in lower means of the progenies than crosses between top $EF_{10\%}$ lines, because the grain yield of the selected $DH_{10\%}$ lines remained far below the performance of the top $EF_{10\%}$. Yet, the usefulness of crosses does not only depend on the mean but also on the variance among their offspring. The higher phenotypic and genetic distances observed between DH lines derived from landraces and the EF lines, on one hand, and those obtained among the EF lines (Figure 1 and 2), on the other, suggest that the progeny of crosses between DH lines derived from landraces and the EF lines might release enhanced genotypic variance compared to crosses within EF [42], [43]. Previous studies confirmed the high usefulness of crosses between elite material and unselected landraces [3] or exotic material [44], and support our optimism that introgression of selected DH lines would allow broadening the genetic base of the EF material without compromising on the performance level. We had expected that landraces would harbor numerous alleles absent from the EF germplasm. Surprisingly, we identified most of specific alleles within the EF. This might result from recent introgressions of Lancaster Sure Crop germplasm into the EF material of the University of Hohenheim (W. Schipprack, personal communication). Second, we also need to take into account some ascertainment bias of the SNP chip towards higher diversity among Lancaster germplasm than among European flint, even after the exclusion of the marker set developed by Syngenta [38]. In addition, the number of population-specific alleles might be underestimated with SNP markers because their biallelic nature will not reveal all allelic variants of a gene. It might be more appropriate to investigate the allelic variation of genes by studying haplotypes. Actually, in agreement with a high degree of recombination within the open-pollinated landraces, the rapid decrease of LD observed in the DH lines derived from landraces in comparison to the EF lines (Table 4) supports a high number of new haplotypes within the DH lines. While the introgression of new haplotypes in elite material disrupts gene combinations with positive epistatic effects [45], it also breaks negative trait associations due to linkage [46] and releases new genotypic variation. Further, the rapid decay of LD together with high genotypic variances and absence of population structure within the populations of DH lines derived from landraces enables high resolution association mapping in such germplasm. Thus, new genes and alleles of agronomic interest might be identified with high precision in the DH lines derived from landraces prior to marker-assisted introgression into the elite material.

CONCLUSIONS

Owing to the large estimates of genotypic variance among the DH lines derived from landraces, individual lines with superior performance for agronomic and morphological traits can be selected and introgressed into the elite material. As suggested by the high phenotypic and genetic distance between the DH lines and the EF lines, the generally lower grain yield and testcross performance of DH lines derived from landraces might be well compensated by a large genotypic variance for these trait in the progenies of crosses with EF lines [3]. Further, the improvement of seed set and other traits related to fitness in the synthetic populations suggest that the DH technique might help in purging detrimental alleles present in landraces, apparently without strongly affecting the phenotypic diversity. Creation of DH lines from landraces shows great promise to broaden and improve the genetic basis of the EF breeding material without necessarily introducing negative agronomic features present in the landraces. Furthermore, the rapid decay of LD together with the high genotypic variances and absence of population structure within the populations of DH lines derived from landraces make these lines an ideal tool for high resolution association mapping.

MATERIALS AND METHODS

Plant material

A set of 132 DH lines was produced by KWS SAAT AG (Einbeck, Germany) from the European maize landraces *Bugard* (DH-BU, n = 36), *Gelber Badischer* (DH-GB, n = 31), and*Schindelmeiser* (DH-SC, n = 65) by a proprietary *in-vivo* haploid induction technique similar to the one described by Röber et al. [23]. Passport and primary descriptors of these landraces can be found in the European Union Maize Landrace Data Base [47]. The landraces *Bugard, Gelber Badischer,* and *Schindelmeiser* were maintained by KWS SAAT AG. Testcross performance of the DH lines derived from the landraces *Gelber Badischer* and *Schindelmeiser* were already reported in a previous study [18]. For comparison of the DH lines from landraces with advanced European breeding material, we evaluated 106 elite flint (EF) inbred lines from the breeding program of the University of Hohenheim and an additional set of 150 elite lines (EL) belonging to other germplasm pools (Table S1). The EL material was not further analyzed in this study except for the statistical analyses of lattice designs. The most important founder lines of the European flint germplasm, *i.e.,* F2, F7, DK105 and EP1 were included in the EF material. Lines F2 and F7 are derived from the French landrace *Lacaune,* DK105 from the German landrace *Gelber Badischer* and EP1 from the Spanish landrace *Lizargarote* [5].

The landraces *Bugard, Gelber Badischer* and *Schindelmeiser* were multiplied in the winter nursery in Chile during the season 2009/2010 by sowing 200 plants and pollinating each row with bulked pollen from the other rows. Synthetic populations of *Bugard* (SYN-BU), *Gelber Badischer* (SYN-GB) and *Schindelmeiser* (SYN-SC) were produced by intermating all DH lines of the respective population. For this purpose, pollen of five (SYN-BU, SYN-GB) or ten (SYN-SC) DH lines was bulked to pollinate one ear of each of the remaining DH lines from the respective landrace. This procedure was repeated until all lines were used once in a pollen bulk. Equal numbers of seeds per pollinated ear were bulked to ensure an equal contribution of each DH line to the synthetics.

Marker assays and their biometric analyses

Genomic DNA from the inbred lines was extracted from pooled leaf tissue samples of five seedlings per genotype using the CTAB method [48]. Each line was genotyped for 56,110 SNPs using the MaizeSNP50 BeadChip (Illumina Inc., San Diego, USA) [49]. Quality control of the SNP marker data was performed according to Strigens et al. [50]. Inbred lines showing more than 2% of heterozygous loci were excluded. Lines and SNP markers with call rates below 0.95, as well as SNP markers with minor allele frequency (MAF) below 5% were excluded from further analysis. To avoid ascertainment bias arising from a subset of SNP marker designed to maximize genetic distances among Stiff-stalk and non-Stiff-stalk material, we further excluded the set of 14,810 SNPs developed by Syngenta [38], [49]. In conclusion, a set of 125 DH lines derived from landraces, a set of 97 EF lines, and 24,572 SNP markers remained for genetic analyses after the quality check.

Number of population-specific SNPs and pair-wise modified Rogers' distances (d_W) [51] were calculated for both the EF lines and DH lines derived from each landrace. Minimum, maximum and average d_W were determined among the lines of the four groups, as well as between the EF lines and individual populations of the DH lines. Nei's total genetic diversity (H_t) was estimated over all loci and lines [52]. Genetic diversity within populations (H_{wi}) was computed for each individual population and averaged to obtain the mean genetic diversity within populations (H_w). Overall genetic differentiation (G_{ST}) was calculated as 1-(H_w/H_t), and population-wise genetic differentiation (G_{STi}) as 1-(H_{wi}/H_t). Linkage disequilibrium (LD) was calculated within each population as r^2 values between all pairs of loci for each chromosome [53]. To characterize the extent of LD in Mbp within each group, r^2 values were binned according to the distance between markers in steps of 0.05 Mbp and averaged over chromosomes. The threshold of r^2 below which LD was considered non-significant was set to 0.1 [54].

Experiment 1 (DH from landraces and elite lines)

The 256 elite inbred lines and 132 DH lines derived from landraces were divided into two sets of 200 entries each. Twelve inbred lines were common to both sets to allow for a combined analysis and adjust for potential differences between the experimental sets. The two sets were evaluated in separate but adjacent field trials laid out as a 20-by-10 alpha design with two replications [55]. Single-row plots of 3 m length with 0.75 m distance between rows were overplanted and later thinned to a final plant density of 10 plants m^{-2}. The trials were conducted in 2010 in five environments in South Germany, contrasting in mean air temperature, altitude, nitrogen supply, and cultivation practice. Eckartsweier, located at an altitude of 141 m a.s.l. in the upper Rhine Valley, with the highest average temperatures (9.9°C), is considered optimal for maize cultivation, whereas Oberer Lindenhof, located at an altitude of 700 m a.s.l. on the Swabian Alb, is a marginal environment for maize growing due to low average temperature (6.6°C). The plants at Oberer Lindenhof were thus harvested already at the eight-leaf stage. Both locations were amended with fertilizer according to usual cultivation practice (150 kg N ha^{-1}). In Hohenheim (400 m a.s.l., 8.8°C), the trials were conducted on a conventionally fertilized field (150 kg N ha^{-1}) and on a nitrogen deficient one (0 kg N ha^{-1}), where only P and K fertilization was kept at optimum. The location Kleinhohenheim (435 m.a.s.l., 8.8°C), adjacent to Hohenheim, was cultivated according to organic farming directives and amended with 50 Mg ha^{-1} organic manure of undetermined N-availability.

Sixteen traits were evaluated on a plot basis for all lines. Emergence was determined as the ratio in percent of emerged plants to sown seeds per plot before thinning. Leaf chlorosis was scored between the four-leaf and six-leaf stage on a 1 (no chlorosis) to 9 (severe chlorosis) scale. Fresh above-ground biomass in g m^{-2} was determined two to four times, depending on the location, between the four-leaf and eight-leaf stage by a non-destructive phenotyping platform described in detail by Montes et al. [56]. Relative growth rates per growing degree days (GDD) were calculated by fitting an exponential growth function to the measured fresh above-ground biomass as described in detail by Strigens et al. [57]. For calculation of GDD, minimal and maximal daily temperatures were obtained from weather stations adjacent to the field trials and base temperature was set to 10°C. Female and male flowering were determined as the GDD from sowing until silk emergence and pollen shedding in more than 50% of the plants, respectively. The anthesis-silking interval was expressed in GDD as the difference between female and male flowering. Plant and ear height in cm were determined at maturity as the approximate distance from the soil to the lowest tassel branch, by placing a level staff in the center of each plot. Ear shank and husk flag leaves were scored on a scale from 1 (absent) to 9 (very pronounced). At physiological maturity (black layer), the ears of five plants from the center of each plot were harvested by hand. To determine ear dry matter concentration, the ears were weighed before and after drying at 60°C to a constant weight. Ear length in cm and ear diameter in mm, number of kernel rows and kernels per row of the primary ear were recorded prior to shelling. Average grain yield in g $plant^{-1}$ and 100 kernel weight in g were determined from bulked seeds of the primary and secondary ears (as far as present) of the five plants.

Experiment 2 (Landraces and synthetic populations)

The three landraces *Bugard*, *Gelber Badischer* and *Schindelmeiser* and the corresponding three synthetic populations produced from the intermated DH lines were evaluated in a randomized complete block design with three replications. The trials were conducted in fields adjacent to Experiment 1 in the same environments except Oberer Lindenhof. Four-row plots of 3 m (Eckartsweier, Hohenheim) or 4 m (Kleinhohenheim) length with 0.75 m between rows were overplanted and later thinned to a final plant density of 9 plants m^{-2}.

Emergence, leaf chlorosis, relative growth rate, female and male flowering, and anthesis-silking interval were determined on a plot basis as in Experiment 1. Plant and ear height, ear shank and husk flag leaves scores were determined on 10 plants from each of the two center rows. Occurrence of common smut (*Ustilago maydis*) and barren stalks was recorded on these 20 plants. The two center rows were harvested at maturity with a combine to determine fresh grain yield. A grain sample of ~500 g was dried at 60°C to a constant weight to determine kernel dry matter content and 100 kernel weight. Grain yield was calculated for a final dry matter content of 85%. Five ears each were harvested by hand from the center of the two remaining outer rows, to measure ear length and diameter, kernel rows, and number of kernels per row.

Statistical analyses

For analysis of the phenotypic data, DH-BU lines, DH-GB lines, DH-SC lines, EF lines and EL lines were considered as five populations. The following model was employed to estimate variance components in Experiment 1:

$$y_{ijklm} = \mu + p_i + g_{ij} + e_k + ge_{ijk} + s_{kl} + r_{klm} + b_{klmn} + \varepsilon_{ijklmn}, \quad (1)$$

where μ is the overall mean, p_i the effect of population i, g_{ij} the effect of inbred line j within population i, e_k is the effect of environment k, ge_{ijk} the interaction between inbred line j within population i and environment k, s_{kl} the effect of set l within environment k, r_{klm} the effect of replication m within trial l, b_{klmn} the effect of incomplete block n within replication m, and ε_{ijklmn} the residual. All effects in Eq. (1) except μ and p_i were considered as random. For ear dry matter content, the sum of GDD from female flowering to harvest was additionally taken as covariate to adjust for different harvest dates. Estimates of the genotypic variance and the variance of genotype×environment interactions were computed within each population by restricted maximum likelihood, using a diagonal variance-covariance structure. Significance of the variance components was determined with the Z-test, assuming normal distribution of variance component estimates. Heterogeneity of residual variance among environments was taken into account and the pooled residual variance was calculated as the average of the individual estimates. Heritabilities (h^2) were calculated within populations on an entry-mean basis, according to Hallauer et al. [30].

For calculation of the adjusted means of the lines, best linear unbiased estimates (BLUEs) were computed by considering μ, p_i, and g_{ij} as fixed effects in Eq. (1) while the remaining effects were considered as random. Differences between populations were tested by using Tukey's honest square difference for unbalanced data sets. To compare the performance of DH and EF lines not only for means, we estimated the predicted response from selection, $\Delta G(a)$, as well as the usefulness criterion [58], $U(a)$, of the populations of DH and elite lines. The parameter $U(a)$ combines the estimate of the population mean and $\Delta G(a)$ and allows a comparison between populations, with regard to the prospects to identify individuals or lines with superior performance. The parameters $\Delta G(a)$ and $U(a)$ were computed following Prigge et al. [27] for a selected proportion of $a = 10\%$ ($U_{10\%}$) and 40% ($U_{40\%}$), corresponding to a selection intensity of $i = 1.76$ and 0.97, respectively. Response from selection was computed as $\Delta G(a) = i(a)h\sigma_g$, where $i(a)$ is the selection intensity, h the square root of the heritability, and σ_g the genotypic standard deviation [59]. The usefulness criterion was calculated as $U(a) = \mu \pm \Delta G(a)$, where μ is the mean of the respective set of lines. The sign of $\Delta G(a)$ was chosen depending on whether higher values of the trait expression were regarded as positive or negative. The best 10% DH lines across the three landraces ($DH_{10\%}$) and EF lines ($EF_{10\%}$) were selected according to the index $IP = 2\times$ear dry matter content+grain yield, commonly used by maize breeders in Central Europe. Within each breeding group, phenotypic correlations (r_p) between grain yield and the remaining traits were determined as Pearson's correlation coefficient and the significance level was Bonferroni-corrected to account for multiple comparisons among populations. Pair-wise Euclidean distances (ED) were calculated among the EF inbred lines and DH lines derived from landraces from the adjusted entry means of the flint genotypes for the sixteen traits evaluated in Experiment 1, centered to mean zero and scaled to unit variance. Minimum, maximum and average of ED were calculated within each population, as well as between the EF lines and individual populations of DH lines.

In Experiment 2, the following model was used in a first step to estimate adjusted means of and test for differences between the six populations: $y_{ijk} = \mu + p_i + e_j + pe_{ij} + r_{jk} + \varepsilon_{ijk}$, **(2)** where μ is the overall mean, p_i the effect of the entry i (landraces and synthetic populations), e_j the effect of environment j, pe_{ij} the interaction between landrace or synthetic population i and environment j, r_{jk} the effect of replication k within environment j, and ε_{ijk} the residual.

All effects except μ and p_i were considered as random in Eq. (2). The genotypic variance within each landrace and each synthetic population was determined for traits measured on a single plant basis by estimating the residual variance among individuals within each landrace and synthetic population. An F-test was performed to evaluate the significance of differences between estimates of the residual variance among individuals. In a second step, the following model was used to test for systematic changes between source landraces and synthetic populations resulting from the use of the DH technique:

$$y_{ijkl} = \mu + g_i + t_j + gt_{ij} + e_k + ge_{ik} + te_{jk} + gte_{ijk} + r_{kl} + \varepsilon_{ijkl}, \quad (3)$$

where μ is the overall mean, g_i the effect of the landrace i (*Bugard, Gelber Badischer, Schindelmeiser*), t_j the effect of the population type j (source landrace vs. synthetic population), e_k the effect of environment k, gt_{ij}, ge_{ik}, te_{jk}, *and* gte_{ijk} the interactions among landrace i, population type j and environment k, r_{kl} the effect of replication l within environment k, and ε_{ijkl} the residual. The parameter μ, g_i and t_j were considered as fixed in Eq. (3).

All calculations were performed within the R-environment [60]. Mixed model analyses were performed using the package ASReml for the R-environment [61].

ACKNOWLEDGMENTS

We are grateful to KWS SAAT AG for kindly providing the DH lines from landraces and to Dr. T. Presterl from KWS SAAT AG for a fruitful collaboration. We are indebted to Dr. E. Orsini for supervising the initial steps of the project and to the technical team of the University of Hohenheim for their dedicated work in the field trials. Further, we thank Prof. Dr. F.H. Utz for his advice on the statistical analysis, J. Muminovic and V. Windhausen for reviewing the manuscript, and D. Neumeister, M. Wohlrab, N. Münch, S. Hütter and V. Klotz, for their appreciated help in phenotyping the material. We also thank the editor and the anonymous reviewers for their valuable suggestions.

Author Contributions

Conceived and designed the experiments: AEM JCR WS. Performed the experiments: AS WS. Analyzed the data: AS. Contributed reagents/materials/analysis tools: WS. Wrote the paper: AS JCR WS AEM.

REFERENCES

1. Rebourg C, Chastanet M, Gouesnard B, Welcker C, Dubreuil P, et al. (2003) Maize introduction into Europe: the history reviewed in the light of molecular data. Theoretical and Applied Genetics 106: 895–903. doi: 10.1007/s00122-002-1140-9 View Article PubMed/NCBI Google Scholar
2. Tenaillon MI, Charcosset A (2011) A European perspective on maize history. Comptes Rendus Biologies 334: 221–228. doi: 10.1016/j.crvi.2010.12.015 View Article PubMed/NCBI Google Scholar
3. Gouesnard B, Dallard J, Bertin P, Boyat A, Charcosset A (2005) European maize landraces: genetic diversity, core collection definition and methodology of use. Maydica 50: 115–234. View Article PubMed/NCBI Google Scholar
4. Barrière Y, Alber D, Dolstra O, Lapierre C, Motto M, et al. (2006) Past and prospects of forage maize breeding in Europe. II. History, germplasm evolution and correlative agronomic changes. Maydica 51: 435–449. View Article PubMed/NCBI Google Scholar
5. Messmer M, Melchinger AE, Boppenmaier J, Herrmann RG, Brunklaus-Jung E (1992) RFLP analyses of early-maturing European maize germ plasm I. Genetic diversity among flint and dent inbreds. Theoretical and Applied Genetics 83: 1003–1012. iew Article PubMed/NCBI Google Scholar
6. Reif JC, Hamrit S, Heckenberger M, Schipprack W, Maurer HP, et al. (2005) Trends in genetic diversity among European maize cultivars and their parental components during the past 50 years. Theoretical and Applied Genetics 111: 838–845. doi: 10.1007/s00122-005-0004-5 View Article PubMed/NCBI Google Scholar
7. Reif JC, Hamrit S, Heckenberger M, Schipprack W, Peter Maurer H, et al. (2005) Genetic structure and diversity of European flint maize populations determined with SSR analyses of individuals and bulks. Theoretical and Applied Genetics 111: 906–913. doi: 10.1007/s00122-005-0016-1 View Article PubMed/NCBI Google Scholar
8. Escholz TW, Peter R, Stamp P, Hund A (2008) Genetic diversity of Swiss maize (Zea mays L. ssp. mays) assessed with individuals and bulks on agarose gels. Genetic Resources and Crop Evolution 55: 971–983. View Article PubMed/NCBI Google Scholar

9. Escholz TW, Stamp P, Peter R, Leipner J, Hund A (2010) Genetic structure and history of Swiss maize (Zea mays L. ssp. mays) landraces. Genetic Resources and Crop Evolution 57: 71–84. doi: 10.1007/s10722-009-9452-0 View Article PubMed/NCBI Google Scholar
10. Gouesnard B, Dallard J, Panouillé A, Boyat A (1997) Classification of French maize populations based on morphological traits. Agronomie 17: 491–498. doi: 10.1007/s10722-009-9452-0 View Article PubMed/NCBI Google Scholar
11. Lucchin M, Barcaccia G, Parrini P (2003) Characterization of a flint maize (Zea mays L. convar. mays) Italian landrace: I. Morpho-phenological and agronomic traits. Genetic Resources and Crop Evolution 5: 315-327. doi: 10.1007/s10722-009-9452-0 View Article PubMed/NCBI Google Scholar
12. Peter R, Escholz TW, Stamp P, Liedgens M (2009) Swiss Flint maize landraces—A rich pool of variability for early vigour in cool environments. Field Crops Research 110: 157–166. doi: 10.1007/s10722-009-9452-0 View Article PubMed/NCBI Google Scholar
13. Peter R, Escholz TW, Stamp P, Liedgens M (2009) Early growth of flint maize landraces under cool conditions. Crop Science 49: 169–178. doi: 10.1007/s10722-009-9452-0 View Article PubMed/NCBI Google Scholar
14. Schneider DN, Freitag NM, Liedgens M, Feil B, Stamp P (2011) Early growth of field-grown Swiss flint maize landraces. Maydica 56: 1702. View Article PubMed/NCBI Google Scholar
15. Rodríguez VM, Romay MC, Ordás A, Revilla P (2010) Evaluation of European maize (Zea mays L.) germplasm under cold conditions. Genetic Resources and Crop Evolution 57: 329–335. View Article PubMed/NCBI Google Scholar
16. Revilla P, Boyat A, Álvarez A, Gouesnard B, Ordás B, et al. (2006) Contribution of autochthonous maize populations for adaptation to European conditions. Euphytica 152: 275–282. View Article PubMed/NCBI Google Scholar
17. Ferro RA, Brichette I, Evgenidis G, Karamaligkas C, Moreno-González J (2006) Variability in European maize (Zea mays L.) landraces under high and low nitrogen inputs. Genetic Resources and Crop Evolution 54: 295–308. View Article PubMed/NCBI Google Scholar
18. Wilde K, Burger H, Prigge V, Presterl T, Schmidt W, et al. (2010) Testcross performance of doubled-haploid lines developed from European flint maize landraces. Plant Breeding 129: 181–185. View Article PubMed/NCBI Google Scholar
19. Malvar RA, Butrón A, Álvarez A, Ordás B, Soengas P, et al. (2004)

Evaluation of the European Union maize landrace core collection for resistance to Sesamia nonagrioides (Lepidoptera: Noctuidae) and Ostrinia nubilalis (Lepidoptera: Crambidae). Journal of Economic Entomology 97: 628–634. doi: 10.1603/0022-0493-97.2.628 View Article PubMed/NCBI Google Scholar

20. Malvar RA, Butrón A, Álvarez A, Padilla G, Cartea M, et al. (2007) Yield performance of the European Union Maize Landrace Core Collection under multiple corn borer infestations. Crop Protection 26: 775–781. doi: 10.1016/j.cropro.2006.07.004 View Article PubMed/NCBI Google Scholar
21. Hoisington D, Khairallah M, Reeves T, Ribaut J-M, Skovmand B, et al. (1999) Plant genetic resources: what can they contribute toward increased crop productivity? Proceedings of the National Academy of Sciences 96: 5937–5943. doi: 10.1016/j.cropro.2006.07.004 View Article PubMed/NCBI Google Scholar
22. Crnokrak P, Barrett SCH (2002) Perspective: purging the genetic load: a review of the experimental evidence. Evolution 56: 2347–2358. doi: 10.1016/j.cropro.2006.07.004 View Article PubMed/NCBI Google Scholar
23. Röber FK, Gordillo GA, Geiger HH (2005) In vivo haploid induction in maize-performance of new inducers and significance of doubled haploid lines in hybrid breeding. Maydica 50: 275–283. doi: 10.1016/j.cropro.2006.07.004 View Article PubMed/NCBI Google Scholar
24. Charlesworth D, Charlesworth B (1992) The effects of selection in the gametophyte stage on mutational load. Evolution 46: 703–720. doi: 10.2307/2409639 View Article PubMed/NCBI Google Scholar
25. Eder J, Chalyk S (2002) In vivo haploid induction in maize. Theoretical and Applied Genetics 104: 703–708 Available: http://www.ncbi.nlm.nih.gov/pubmed/12582677. View Article PubMed/NCBI Google Scholar
26. Gallais A (1990) Quantitative genetics of doubled haploid populations and application to the theory of line development. Genetics 4: 199–206. doi: 10.2307/2409639
27. Prigge V, Babu R, Das B, Rodriguez MH, Atlin GN, et al. (2012) Doubled haploids in tropical maize: II. Quantitative genetic parameters for testcross performance. Euphytica 185: 453–463. doi: 10.2307/2409639
28. Flint-Garcia SA, Buckler ES, Tiffin P, Ersoz E, Springer NM (2009) Heterosis is prevalent for multiple traits in diverse maize germplasm. PloS one 4: e7433. doi: 10.1371/journal.pone.0007433
29. Schnell FW (1959) Mais. In: Rudorf W, editor. Dreißig Jahre Züchtungsforschung. Stuttgart: Fischer Verlag. p. 140–141.

30. Hallauer A, Carena MJ, Miranda JB (2010) Quantitative genetics in maize breeding. New York, NY: Springer Science and Business Media LLC.

31. Murigneux A, Baud S, Beckert M (1993) Molecular and morphological evaluation of doubled-haploid lines in maize. 2. Comparison with single-seed-descent lines. Theoretical and Applied Genetics 87: 278–287. doi: 10.1007/bf00223777

32. Lashermes P, Gaillard A, Beckert M (1988) Gynogenetic haploid plants analysis for agronomic and enzymatic markers in maize (Zea mays L.). Theoretical and Applied Genetics 76: 570–572. doi: 10.1007/bf00223777

33. Martin M, Miedaner T, Schwegler DD, Kessel B, Ouzunova M, et al. (2012) Comparative Quantitative Trait Loci Mapping for Gibberella Ear Rot Resistance and Reduced Deoxynivalenol Contamination across Connected Maize Populations. Crop Science 52: 32–43. doi: 10.1007/bf00223777

34. Edmeades G, Bolanos J, Hernandez M, Bello S (1993) Causes for silk delay in a lowland tropical maize population. Crop Science 33: 1029–1035. doi: 10.2135/cropsci1993.0011183x003300050031x

35. Betran F, Beck D, Bänziger M, Edmeades G (2003) Secondary traits in parental inbreds and hybrids under stress and non-stress environments in tropical maize. Field Crops Research 83: 51–65. doi: 10.2135/cropsci1993.0011183x003300050031x

36. Melchinger AE, Geiger HH, Schnell F (1985) Reciprocal differences in single-cross hybrids and their F2 and backcross progenies in maize. Maydica 30: 395–405. doi: 10.2135/cropsci1993.0011183x003300050031x

37. Van Inghelandt D, Melchinger AE, Lebreton C, Stich B (2010) Population structure and genetic diversity in a commercial maize breeding program assessed with SSR and SNP markers. Theoretical and Applied Genetics 120: 1289–1299. doi: 10.1007/s00122-009-1256-2

38. Frascaroli E, Schrag TA, Melchinger AE (2012) Genetic diversity analysis of elite European maize (Zea mays L.) inbred lines using AFLP, SSR, and SNP markers reveals ascertainment bias for a subset of SNPs. Theoretical and Applied Genetics In press. doi: 10.2135/cropsci1993.0011183x003300050031x

39. Lauer S, Hall BD, Mulaosmanovic E, Anderson SR, Nelson BK, et al. (2012) Morphological changes in parental lines of Pioneer brand maize hybrids in the U. S. Central Corn Belt. Crop Science 52: 1033–1043.

40. Duvick DN (2005) Genetic progress in yield of United States Maize (Zea mays L.). Maydica 50: 193–202.

41. Troyer A F, Wellin EJ (2009) Heterosis decreasing in hybrids: Yield test inbreds. Crop Science 49: 1969–1976.

42. Hung H-Y, Browne C, Guill K, Coles N, Eller M, et al. (2012) The relationship between parental genetic or phenotypic divergence and progeny variation in the maize nested association mapping population. Heredity 108: 490–499. doi: 10.1038/hdy.2011.103

43. Melchinger A (1999) Genetic diversity and heterosis. In: Coors JG, Pandey S, editors. The genetics and exploitation of heterosis in crops. Madison, WI: ASA, CSSA, and SSSA. pp. 99–118.

44. Crossa J, Gardner C (1987) Introgression of an exotic germplasm for improving an adapted maize population. Crop Science 27: 187–190. doi: 10.2135/cropsci1987.0011183x002700020008x

45. Melchinger AE, Geiger HH, Utz HF, Schnell F (2003) Effect of recombination in the parent populations on the means and combining ability variances in hybrid populations of maize (Zea mays L.). Theoretical and Applied Genetics 106: 332–340. doi: 10.1007/s00122-002-1000-7

46. Gallais A (1990) Théorie de la sélection en amélioration des plantes. Paris: Masson.

47. European Maize Landraces Data Base. Available: http://www.ensam.inra.fr/gap/BD/Eumldb.zip. Accessed 2013 Jan 29.

48. CIMMYT (2005) Laboratory protocols: CIMMYT applied molecular genetics laboratory. 3rd ed. CIMMYT.

49. Ganal MW, Durstewitz G, Polley A, Bérard A, Buckler ES, et al. (2011) A large maize (Zea mays L.) SNP genotyping array: development and germplasm genotyping, and genetic mapping to compare with the B73 reference genome. PloS one 6: e28334. doi: 10.1371/journal.pone.0028334

50. Strigens A, Freitag NM, Gilbert X, Grieder C, Riedelsheimer C, et al. (2012) Association mapping for chilling tolerance in elite flint and dent maize inbred lines evaluated in growth chamber and field experiments. Plant, Cell and Environment In review. doi: 10.1111/pce.12096

51. Wright S (1978) Evolution and genetics of populations, variability within and among natural populations, vol 4. Chicago: The University of Chicago Press.

52. Nei M (1977) F-statistics and analysis of gene diversity in subdivided populations. Annals of Human Genetics 41: 225–233. doi: 10.1111/j.1469-1809.1977.tb01918.x

53. Hill WG, Robertson A (1966) Linkage disequilibrium in finite populations. Theoretical and Applied Genetics 38: 226–231. doi: 10.1111/pce.12096

54. Zhu C, Gore MA, Buckler ES, Yu J (2008) Status and prospects of association mapping in plants. The Plant Genome Journal 1: 5–20. doi: 10.3835/plantgenome2008.02.0089

55. Patterson HD, Williams E (1976) A new class of resolvable incomplete block designs. Biometrika 63: 83–92. doi: 10.3835/plantgenome2008.02.0089

56. Montes JM, Technow F, Dhillon BS, Mauch F, Melchinger AE (2011) High-throughput non-destructive biomass determination during early plant development in maize under field conditions. Field Crops Research 121: 268–273. doi: 10.3835/plantgenome2008.02.0089

57. Strigens A, Grieder C, Haussmann BI, Melchinger AE (2012) Genetic variation among inbred lines and testcrosses of maize for early growth parameters and their relationship to final dry matter yield. Crop Science 52: 1084–1092. doi: 10.3835/plantgenome2008.02.0089

58. Schnell F (1983) Probleme der Elternwahl - ein Überblick. Arbeitstagung der Vereinigung der Saatzuchtleiter in Gumpenstein, Austria, 22.–24. Nov. Gumpenstein, Austria: Verlag und Druck der Bundesanstalt für alpenländische Landwirtschaft.

59. Falconer D, Mackay TFC (1996) Introduction to quantitative genetics. 4th ed. New York: Longmann Scientific & Technical.

60. R Development Core Team (2011) R: a language and environment for statistical computing.

61. Butler D, Cullis B, Gilmour A, Gogel B (2007) Analysis of mixed models for S language environments. ASReml-R reference manual. 2.0. Brisbane: The State of Queensland, Department of Primary Industries and Fisheries.

Chapter 2

MAPPING GENETIC DIVERSITY OF CHERIMOYA (ANNONA CHERIMOLA MILL.): APPLICATION OF SPATIAL ANALYSIS FOR CONSERVATION AND USE OF PLANT GENETIC RESOURCES

[1,2]Maarten van Zonneveld, [1]Xavier Scheldeman, [3]Pilar Escribano, [3]María A. Viruel, [2,4]Patrick Van Damme, [5]Willman Garcia, [6]César Tapia, [7]José Romero, [8]Manuel Sigueñas, [3]José I. Hormaza

[1,2]Maarten van Zonneveld, [1]Xavier Scheldeman, [3]Pilar Escribano, [3]María A. Viruel, [2,4]Patrick Van Damme, [5]Willman Garcia, [6]César Tapia, [7]José Romero, [8]Manuel Sigueñas, [3]José I. Hormaza

[1]Bioversity International, Regional Office for the Americas, Cali, Colombia

[2]Ghent University, Faculty of Bioscience Engineering, Gent, Belgium

[3]Instituto de Hortofruticultura Subtropical y Mediterránea, (IHSM-UMA-CSIC), Estación Experimental La Mayora, Algarrobo-Costa, Málaga, Spain

[4]World Agroforestry Centre (ICRAF), GRP1 - Domestication, Nairobi, Kenya

[5]PROINPA, Oficina Regional Valle Norte, Cochabamba, Bolivia

[6]Instituto Nacional Autónomo de Investigaciones Agropecuarias (INIAP)

Panamericana sur km1, Quito, Ecuador
[7]Naturaleza y Cultura Internacional (NCI), Loja, Ecuador
[8]Instituto Nacional de Innovación Agrícola (INIA), La Molina, Lima, Peru

ABSTRACT

There is a growing call for inventories that evaluate geographic patterns in diversity of plant genetic resources maintained on farm and in species' natural populations in order to enhance their use and conservation. Such evaluations are relevant for useful tropical and subtropical tree species, as many of these species are still undomesticated, or in incipient stages of domestication and local populations can offer yet-unknown traits of high value to further domestication. For many outcrossing species, such as most trees, inbreeding depression can be an issue, and genetic diversity is important to sustain local production. Diversity is also crucial for species to adapt to environmental changes. This paper explores the possibilities of incorporating molecular marker data into Geographic Information Systems (GIS) to allow visualization and better understanding of spatial patterns of genetic diversity as a key input to optimize conservation and use of plant genetic resources, based on a case study of cherimoya (Annona cherimola Mill.), a Neotropical fruit tree species. We present spatial analyses to (1) improve the understanding of spatial distribution of genetic diversity of cherimoya natural stands and cultivated trees in Ecuador, Bolivia and Peru based on microsatellite molecular markers (SSRs); and (2) formulate optimal conservation strategies by revealing priority areas for in situ conservation, and identifying existing diversity gaps in ex situ collections. We found high levels of allelic richness, locally common alleles and expected heterozygosity in cherimoya's putative centre of origin, southern Ecuador and northern Peru, whereas levels of diversity in southern Peru and especially in Bolivia were significantly lower. The application of GIS on a large microsatellite dataset allows a more detailed prioritization of areas for in situ conservation and targeted collection across the Andean distribution range of cherimoya than previous studies could do, i.e. at province and department level in Ecuador and Peru, respectively.

INTRODUCTION

Many useful tropical and subtropical tree species, even those commonly cultivated, are still in incipient stages of domestication, with their genetic resources often principally or exclusively, present in situ, i.e. on farm in home gardens or orchards and/or in natural populations. The local diversity of these tree species could offer yet-unknown traits of high value to further domestication [1]. For many outcrossing species, such as most tropical tree species, this genetic diversity is important to sustain local production as many of these species are vulnerable to inbreeding depression [2]. Diversity is also a key factor for adaption to environmental changes [2]. However, tree species are increasingly vulnerable to losses of genetic diversity, referred to as genetic erosion, due to decreased population sizes resulting from land use changes and land degradation, and due to changes in local climate that may select against some genotypes [3]. Therefore, there is a growing call to assess the conservation status of the genetic resources of tree species [4].

The formulation of effective and efficient conservation strategies requires a thorough understanding of spatial patterns of genetic diversity [5]. A better knowledge of areas of high genetic diversity is also important in optimizing the use of genetic resources, as the likelihood to find interesting materials for breeding is higher where levels of genetic diversity are maximal [6], [7]. Initiatives to prioritize research on global plant genetic resources, such as those lead by the Food and Agriculture Organization of the United Nations (FAO), include calls for more inventories and surveys to increase understanding of variation in plant genetic resources, explicitly referring to the application of molecular tools in such assessments [8], [9].

This study focuses on cherimoya (Annona cherimola Mill.), an underutilized fruit tree species that belongs to the Annonaceae, a family included within the Magnoliales in the Eumagnoliid clade among the early-divergent angiosperms [10]. This Neotropical tree species still is in its initial stages of domestication [11] and it is considered at high risk of losing valuable genetic material from its genepool [12]. Cherimoya fruits are widely praised for their excellent organoleptic characteristics, and the species is therefore considered to have a high potential for commercial production and income generation for both small and large-scale producers in subtropical climates [13]. Cherimoya presents protogynous dichogamy, i.e. it has hermaphroditic flowers wherein female and male parts do not mature simultaneously, which favors outcrossing in its native range [14]. For commercial production, hand pollination with pollen and stamens is common practice due to lack in overlap of the female and male stages and absence of pollinating agents outside its native range [14]. At present, advanced commercial production is found in Spain, the world's largest cherimoya producer, with around 3000 ha of plantations, while small-scale cultivation occurs throughout the Andes, Central America and Mexico.

Most early chroniclers and scientists proposed the Andean region, and more specifically, the valleys of southern Ecuador and northern Peru, as cherimoya's centre of origin [12], [15], [16]. The existence of natural cherimoya forest patches, which are scattered across the inter-Andean valleys in Ecuador and northern Peru, supports this hypothesis. Nonetheless the possibility that these are feral populations cannot be excluded. This phenomenon has been observed in the case of several fruit tree species, such as olives [17]. An alternative hypothesis for the centre of origin of cherimoya is Central America [18], which would imply that the area of northern Peru and southern Ecuador is a secondary centre of diversity. Most relatives of cherimoya are native to Central America and southern Mexico, which is an argument in favor of this alternate hypothesis (H. Rainer, Institute of Botany, University of Vienna, 2011, pers. comm.). In any case, cherimoya fruits were consumed in the Andean region in antiquity [12] and the movement of germplasm across Mesoamerica, southern Mexico and the Andes probably took place in pre-Columbian times.

The conservation status of cherimoya genetic resources has improved considerably in recent years. Due to increasing commercial prices for cherimoya at local markets, Andean farmers are stimulated to conserve in situ the cherimoya trees growing in their backyards. Indeed, trees established in home gardens and orchards are common throughout the Andean region in Bolivia, Ecuador and Peru, which usually originate from planted local seeds or chance seedlings [11], and among them some individuals show promising traits for future breeding programs [19]. In Peru, the local cultivar 'Cumbe' is already fetching retail prices significantly above the prices of unselected cherimoya fruit types [20]. In contrast to most tropical and subtropical underutilized fruit tree species, cherimoya genetic resources are also well conserved ex situ. Several field collections have been established in Spain, Peru and Ecuador, preserving over 500 different accessions [11], [21]. The Spanish collection based at la Estación Experimental La Mayora in Malaga, which holds over 300 accessions (190 of them collected in the Andean region), is currently used as a source of materials for the Spanish cherimoya breeding program and has been thoroughly analyzed using isozymes [22]-[24] and microsatellite markers [11], [25]-[27].

The recent development of new molecular tools in combination with new spatial methods and increased computer capacity has created opportunities for new applications of genetic diversity analyses [28]-[30]. Whereas neutral molecular markers are considered a sound tool to measure patterns and trends in the use and conservation of plant genetic resources [31], Geographic Information Systems (GIS) provide opportunities to carry out spatial analyses of genetic diversity patterns identified with these markers [32]. GIS can be used to interpolate genetic parameters between sampled populations (e.g. [33]-[35]), to apply re-sampling of georeferenced samples within a defined buffer zone [36], [37], or to develop grid-based genetic distance models [38], [39]. GIS are also an acknowledged tool to prioritize areas for conservation of plant genetic resources [40]. Several studies have used spatial analysis to develop conservation strategies for plant genetic resources based on molecular marker data (e.g. [36], [41]). Moreover, results obtained using GIS can be presented in a clear way on maps, which facilitates the incorporation of these findings into the formulation of conservation strategies and the implementation of conservation measures [42].

In this article we further explore the possibilities of incorporating molecular marker data into GIS to better visualize and understand spatial patterns of genetic diversity, as a key input to optimize conservation and enhance use of local plant diversity, based on a case study of cherimoya. The specific objectives of this article are to (1) apply innovative spatial analysis to improve understanding of the geographic distribution of cherimoya 's genetic diversity in its putative native range, based on microsatellite molecular markers (SSRs); and (2) formulate optimal conservation strategies by prioritizing areas for in situ conservation and identifying existing diversity gaps in ex situ collections. Based on the outcomes, we discuss how these spatial analyses can be used to define possible strategies that guarantee the long term conservation of cherimoya genetic resources and how these analyses can be applied to improve conservation and use of tree and crop genetic resources in general.

Results

A total of 1504 trees were analyzed in this study, i.e. 395 from Bolivia, 351 from Ecuador and 758 from Peru. Of those, 502 are currently conserved in ex situ collections (either in Ecuador, Peru or Spain) whereas the remainder trees were sampled in situ. The molecular analysis included a core set of nine microsatellite loci [27] resulting in 71 different alleles. In all analyses of α-diversity and β-diversity (also referred to as divergence) we applied circular neighborhood re-sampling technique resulting in a total dataset of 48,128 trees (Figure 1). This technique facilitates analysis of patterns in genetic variation across extensive distribution ranges while maintaining high-resolution grids.

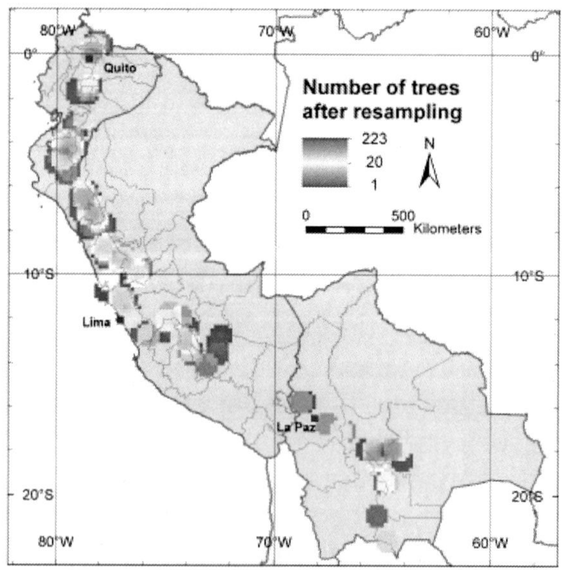

Figure 1. Number of trees per grid after re-sampling.

This map is made with a 10-minutes grid applying a one-degree circular neighborhood.

doi:10.1371/journal.pone.0029845.g001

Allelic richness

Allelic richness is a straightforward measure of genetic diversity that is commonly used in studies based on molecular markers that aim at selecting populations for conservation [5], [43]. Figure 2 presents the distribution of the average number of alleles per locus found in the study area. It clearly shows that a higher number of alleles is present in the northern part of the study area, specifically in northern Peru, around Cajamarca Department, while other areas of high diversity are located on the border zone between Ecuador (Loja Province) and Peru (Piura Department), in the northern part of Ecuador around its capital Quito and in the northern part of the Lima Department in Peru.

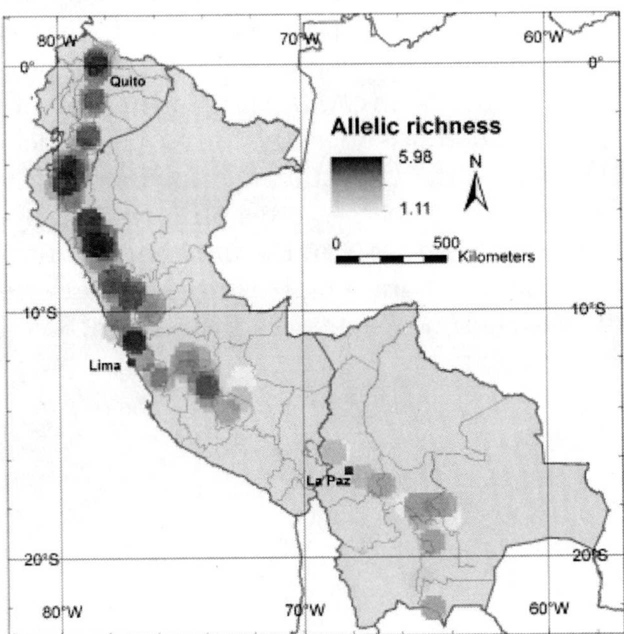

Figure 2. Allelic richness.

This map shows the average number of alleles per locus in all 10-minutes grid cells applying a one-degree circular neighborhood.

doi:10.1371/journal.pone.0029845.g002

Despite the effort to implement a similar sampling density throughout the study area, some areas (often locations with a higher abundance of traditionally managed cherimoya trees and stands) have been sampled more intensively than others (Figure 1), generating a sampling bias [44]. The rarefaction methodology corrects this sampling bias by recalculating allelic richness in each grid cell to a minimum sample size [5]. Figure 3 shows only the grid cells where 20 or more trees were present after applying a one-degree circular neighborhood, and for which allelic richness was corrected following the rarefaction methodology to a minimum sample size of 20 trees. The Cajamarca Department in northern Peru remains the area with the highest diversity, up to an average of 5.18 different alleles per locus.

After correction by rarefaction, diversity in Ecuador, especially around Quito, is reduced, whereas the same seems to happen in the northern part of the Lima Department, in Peru, indicating the presence of a sampling bias around the capitals of both countries. The area around the Peruvian capital Lima, an important commercial cherimoya cultivation area, shows the lowest allelic richness within Peru, probably due to the widespread cultivation of a vegetatively propagated cultivar, 'Cumbe'. Another striking result is that allelic richness in Bolivia, already low in the uncorrected analysis, is even lower with correction of sampling bias, resulting in an even higher contrast between cherimoya genetic diversity in Bolivia and that found in Peru and Ecuador.

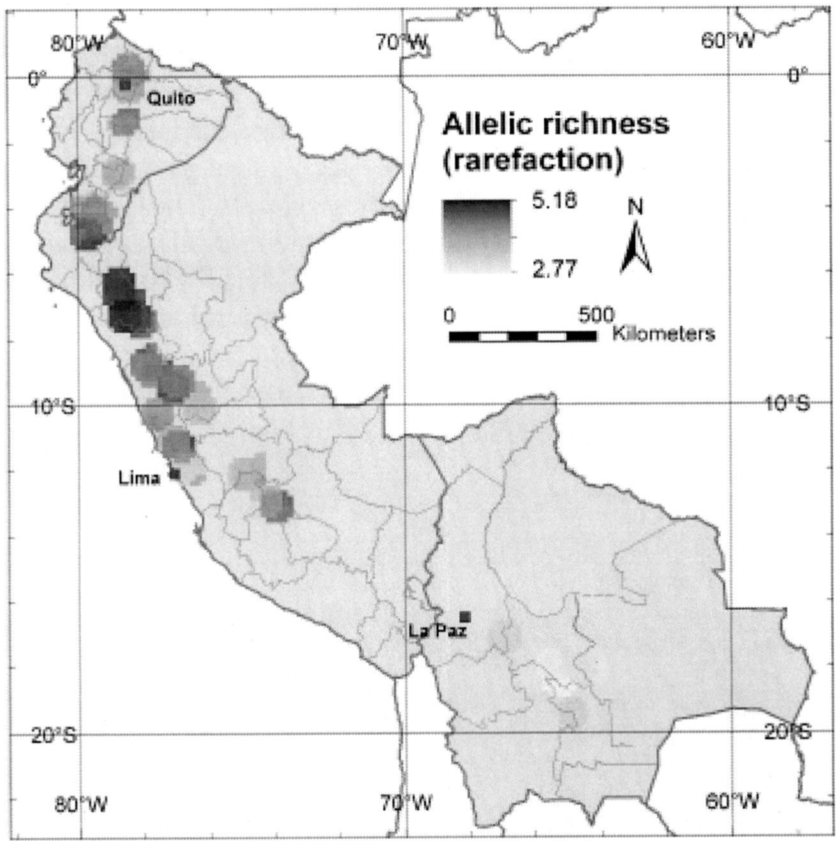

Figure 3. Allelic richness corrected for sample size by using rarefaction.

This map shows the average number of alleles per locus in the 10-minutes grid cells applying a one-degree circular neighborhood and a correction by rarefaction to a minimum sample size of 20 trees.

doi:10.1371/journal.pone.0029845.g003

Locally common alleles

Priority for conservation should be given to populations that retain locally common alleles; these are alleles that occur in high frequency in a limited area, and can indicate the presence of genotypes adapted to specific environments [43]. Figure 4 shows the richness of locally common alleles per locus in the study area. The high diversity levels found in the Cajamarca Department in northern Peru are reconfirmed. Besides harboring the highest number of different alleles, it also contains the highest number of locally common alleles. This makes this area a priority for in situ conservation, both of cultivated trees on farm and of natural stands. The border region between Peru and Ecuador (Piura Department and Loja Province) is another area where a high concentration of locally common alleles has been observed and may, therefore, be a second area to prioritize in situ conservation efforts. To a lesser extent, the area around Quito in Ecuador and the northern part of the Lima Department in Peru also present locally common alleles.

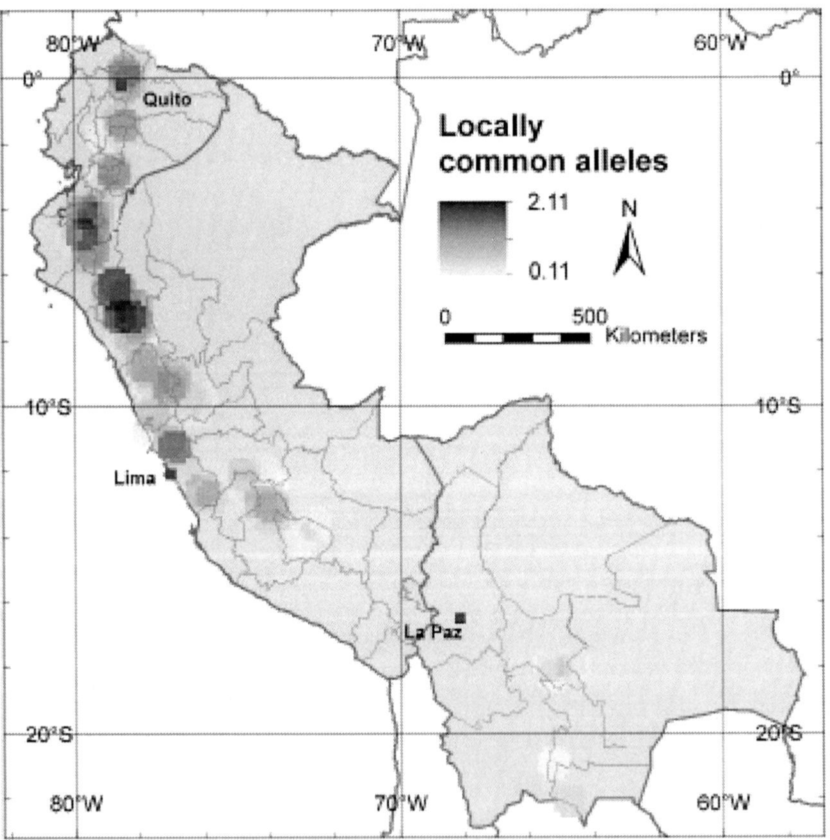

Figure 4. Locally common alleles.

This map shows the average number of alleles per locus in a 10-minutes grid cell that are relatively common (occurring with a frequency higher that 5%) in a limited area (in 25% or less of the grid cells) applying a one-degree circular neighborhood.

doi:10.1371/journal.pone.0029845.g004

Expected Heterozygosity (He), Fixation Index (F) and Genetic Distance (GD)

In situ conservation should focus on viable populations, where inbreeding and subsequent loss of alleles are minimal. Parameters that allow assessment of inbreeding are expected heterozygosity (He) and the fixation index (F). The fixation index (F) was used to detect areas subjected to high inbreeding depression and, as the inverse to that, excess in heterozygosity [45]. Figure 5 shows the values for He in the study area, again confirming Cajamarca Department in northern Peru as the area with the highest genetic diversity. High He values, however, radiate towards the south (as opposed to the higher diversity towards the north found in the allelic richness analyses) indicating higher levels of diversity in terms of heterozygosity in central Peru compared to Ecuador. Figure 6 shows the values for the fixation index, with F values close to 0 in the Cajamarca Department indicating that natural and cultivated cherimoya tree stands in this area have not experienced inbreeding. The highest values for F are observed in central Ecuador, suggesting that the level of inbreeding is highest in that part of cherimoya's Andean distribution range.

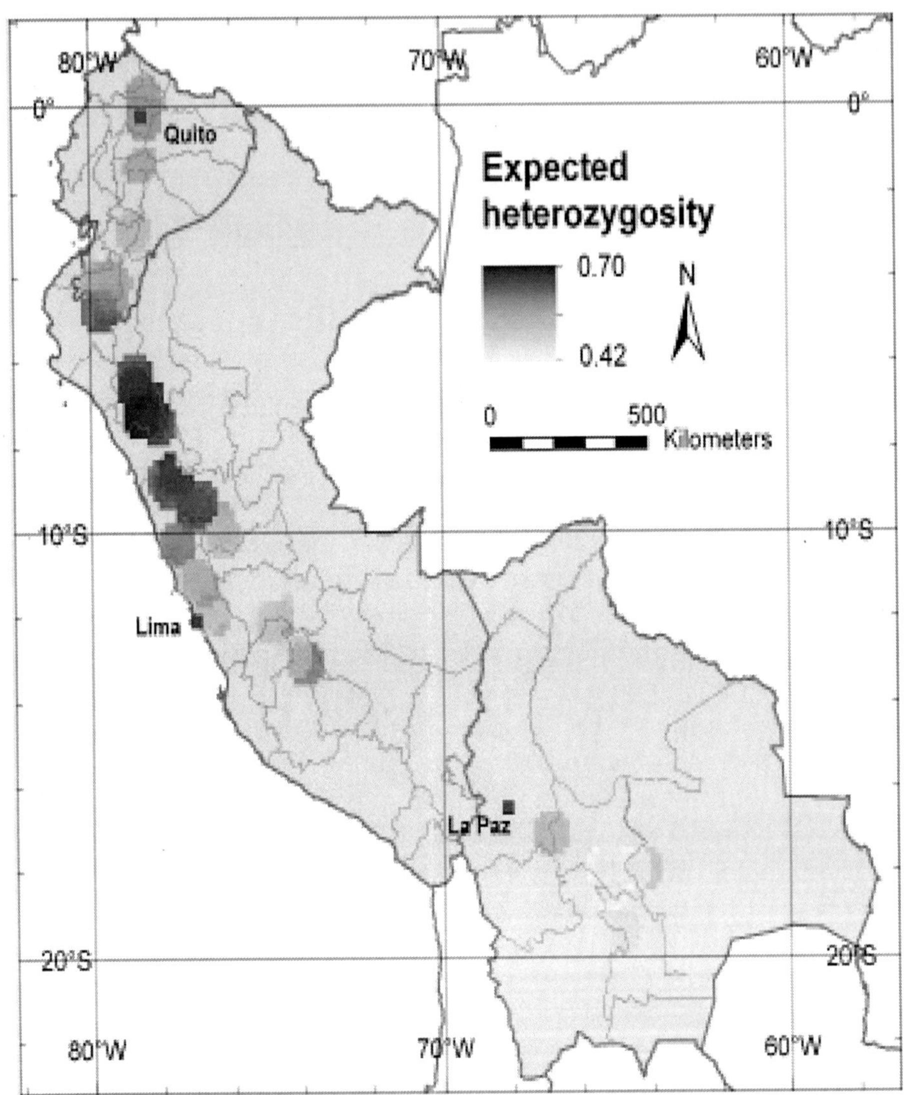

Figure 5. Expected heterozygosity (He).

This map shows the average He value in each 10-minutes grid cell with 20 or more trees applying a one-degree circular neighborhood.

doi:10.1371/journal.pone.0029845.g005

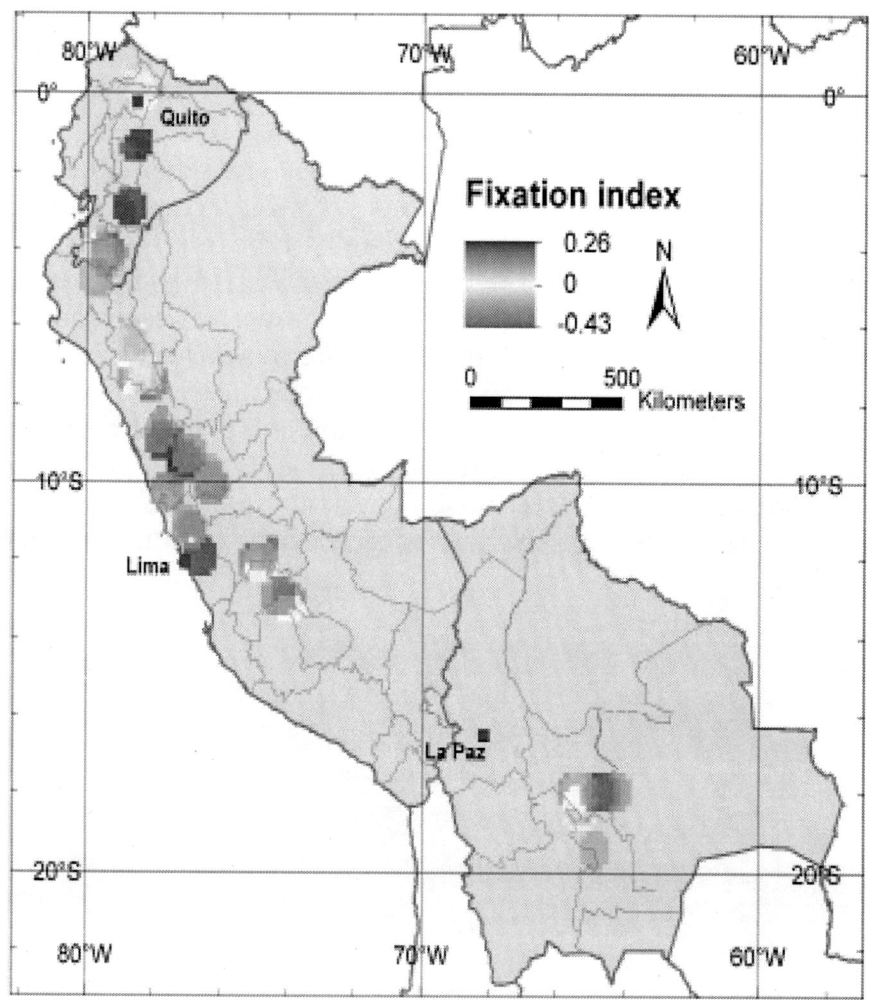

Figure 6. Fixation index (F).

This map shows the average F value in each 10-minutes cell with 20 or more trees applying a one-degree circular neighborhood. Yellow areas indicate cherimoya stands where observed heterozygosity is as expected, red areas indicate stands where observed heterozygosity is lower than expected (indicating inbreeding) whereas observed heterozygosity is higher than expected in green areas.

doi:10.1371/journal.pone.0029845.g006

The most important Peruvian commercial cherimoya cultivation area, located near the Capital Lima, particularly shows negative F values, i.e. excess of heterozygosity. Most of the cherimoyas cultivated in this area are vegetatively propagated clones of the cultivar 'Cumbe' which resulted in highly heterozygous values from the molecular analysis, i.e. the 'Cumbe' accession conserved in the Spanish genebank is heterozygote for eight of the nine microsatellite loci analyzed in this study (Ho value of 0.89). An analysis of the average genetic distance, between the 'Cumbe' accession and the genotypes in each grid cell with 20 or more re-sampled trees in the study area, clearly shows lowest genetic distance values near the Peruvian capital, Lima, indicating that the cherimoya trees in this area are very similar to the cultivar 'Cumbe' (Figure 7). This area clearly differs from the rest of the cherimoya distribution area in our study, which is more likely to be a product of natural gene flow patterns.

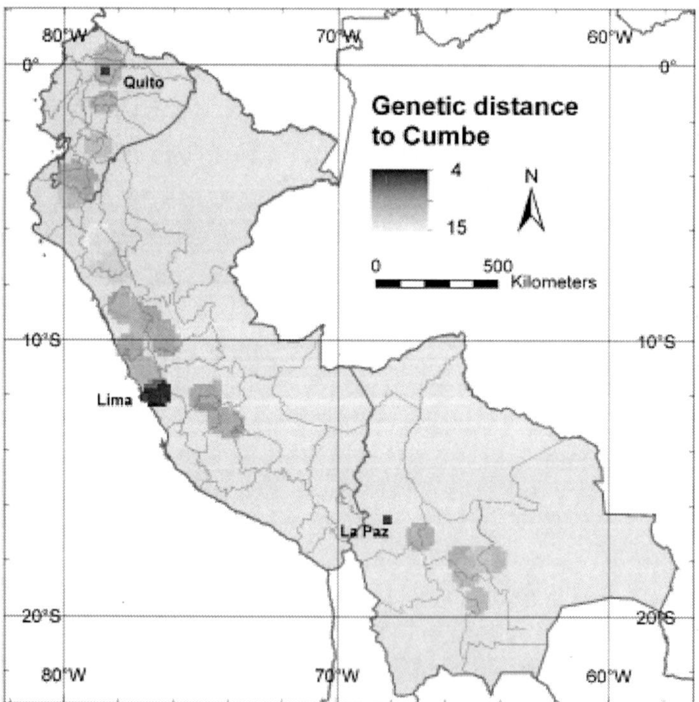

Figure 7. Genetic distance to the Peruvian cultivar 'Cumbe'.

This maps shows the average genetic distance (GD) to the cultivar 'Cumbe', in each 10-minutes cell with 20 or more trees applying a one-degree circular neighborhood. As reference of the cultivar, the 'Cumbe' accession from the collection la Mayora, Malaga, Spain, was used.

doi:10.1371/journal.pone.0029845.g007

β-diversity (divergence)

Besides α-diversity parameters, aimed at identifying those areas with highest allelic richness and balanced allele frequencies, in situ conservation also needs to take into account allelic composition (β-diversity or divergence) as it is possible that populations with low allelic richness possess unique allele compositions, different from those of populations in other areas of the range, which would warrant their in situ conservation [5]. Applying the Structure software (see [46]) and using the statistic parameter ΔK [47] to define the number of clusters with genetically similar trees present in the study area, we differentiated two main populations. Figure 8 shows the differentiation of the populations among distribution areas in cluster A and B, respectively. Cluster A has the highest presence in the areas previously identified as those with the highest allelic richness (Cajamarca Department in northern Peru, border zone between Ecuador and Peru and the area around Quito in Ecuador), whereas cluster B is mainly confined to southern Peru and Bolivia. Bolivian cherimoya trees are almost exclusively assigned to cluster B. Particular areas that did not show a strong linkage to either of the two clusters included the surroundings of the city of Lima and Loja Province in southern Ecuador.

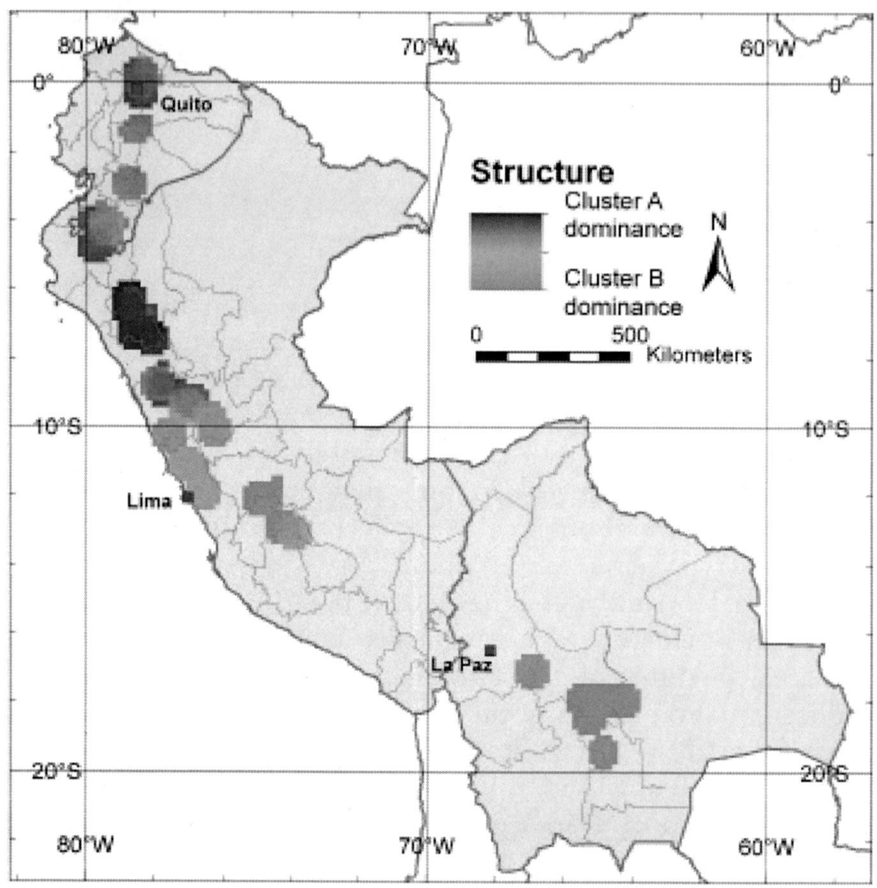

Figure 8. Genetic structure of Andean cherimoya distribution in Population clusters A and B.

This map shows in each 10-minutes cell with 20 or more trees applying a one-degree circular neighborhood, the average probability of finding a cherimoya tree belonging to cluster A or B. Dark blue areas show a higher probability of finding trees belonging to cluster A whereas dark green areas show a higher probability of finding trees belonging to cluster B. Light blue colored areas are not clearly assigned to any of the two clusters.

doi:10.1371/journal.pone.0029845.g008

Ex situ conservation status

Of the 1504 trees included in this study, 502 genotypes are currently conserved in ex situ collections (either in Ecuador, Peru or Spain). Only eight alleles, corresponding to 11% of the total of 71 alleles that have been found in the study area, are not represented in any accession of these collections. Figure 9 shows the distribution of the missing alleles. There is only a small area with a significant portion of missing alleles (3 in total), i.e. in southern Ecuador (Azuay Province). Natural cherimoya forest patches and areas of traditional cherimoya cultivation in this province should be prioritized for future cherimoya collection missions. With almost 90% of alleles found to be present in ex situ collections, it can be concluded that, in general, cherimoya diversity from the countries analyzed is fairly well conserved *ex situ*.

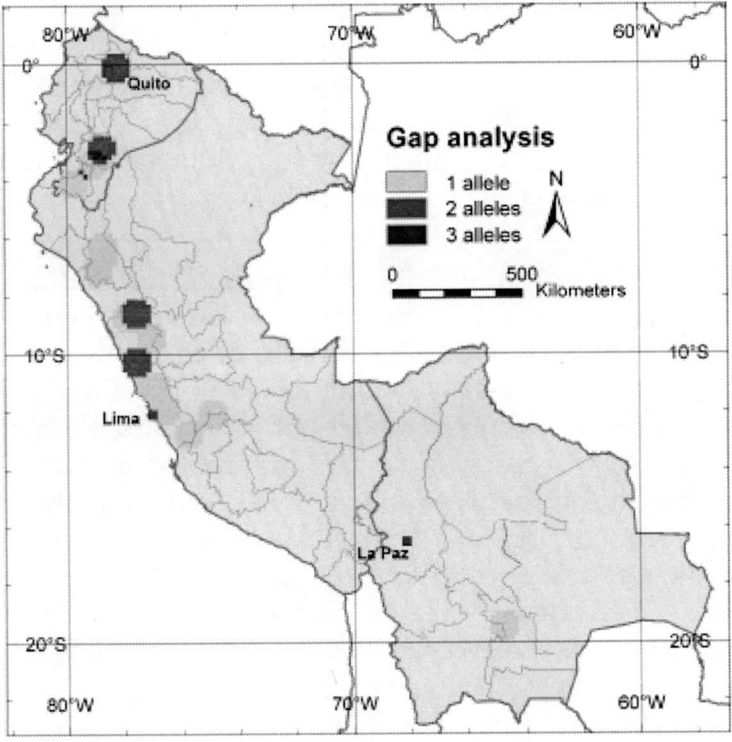

Figure 9. Gap analysis of alleles not found in ex situ collections.

Richness analysis of alleles (eight alleles out of the total of 71 observed alleles) that are not found in any ex situ collection based on 10-minutes grid with a one-degree circular neighborhood.

doi:10.1371/journal.pone.0029845.g009

Distribution range of cherimoya in the Andes

The above results and subsequent conclusions are obviously only of practical use if the sampling performed was indeed representative for the distribution of cherimoya in the study area. Maxent species distribution modeling software was applied to model cherimoya's distribution range in Ecuador, Peru and Bolivia based on the climatic niche in which the 1504 sampled trees of our study were located. The modeled distribution was then compared with the sampled areas in these countries.

Cross-validation, to evaluate the quality of the distribution model, returned an Area Under Curve (AUC) value of 0.9, which indicates good model performance [48]. AUC is a commonly used parameter in the validation of distribution models. Another measure of validation, the Kappa value, returned a value of 0.799 indicating the model performed even excellent [49].

In general, sampling covered most of the cherimoya-modeled distribution (Figure 10); 46% of the modeled distribution area is covered by grid cells with 20 or more re-sampled trees (Figure 10, dark blue areas). In 24.5% of the potential area of cherimoya occurrence less than 20 trees were re-sampled (light blue areas) whereas 29.5% of the modeled range was not sampled (red areas) and are considered sample gaps. The largest sample gaps are located in northern Peru in the transition zone between the Peruvian Andes and the Amazon (in the Departments of San Martin and Amazonas) and in southern Peru (in the Departments of Junín, Pasco, Huancavelica, Ayacucho and Puno). The Andean-Amazon transition zone should be priority for future complementary cherimoya collection trips because it is adjacent to an area where already high levels of diversity have been found, i.e. Cajamarca Department in northern Peru.

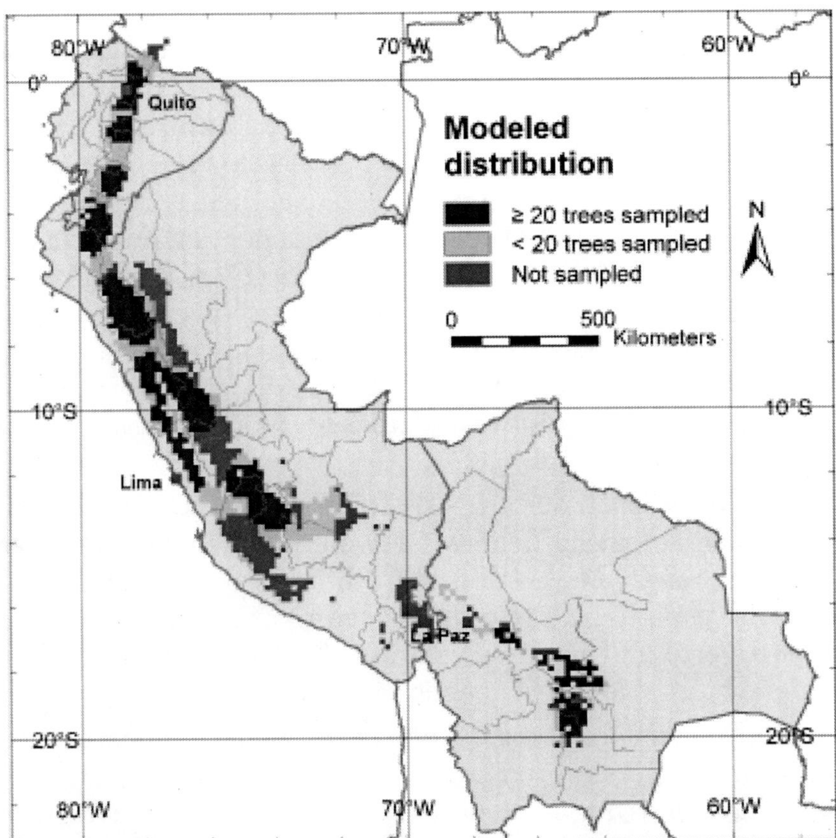

Figure 10. Modeled distribution of cherimoya.

Areas of the modeled distribution in dark blue are covered by the 10-minutes grid cells with 20 or more trees applying circular neighborhood. Light blue areas of modeled distribution coincide with grid cells that contain less than 20 trees after re-sampling. Red areas indicate potential areas for cherimoya occurrence and cultivation that have not been in sampled.

doi:10.1371/journal.pone.0029845.g010

Cherimoya was predicted absent by the distribution model in a significant area of southern Peru, indicating that the environmental conditions in substantial parts of that region are not suitable for cherimoya cultivation (Figure 10). This explains why no trees have been sampled in that area.

DISCUSSION

Areas of high diversity in the cherimoya centre of origin

Our results are in line with a previous genetic study of the Spanish cherimoya collection that also distinguished populations in Ecuador and northern Peru from those in southern Peru [11], and corroborate with results from isozyme markers that showed high genetic variation present in Peru and Ecuador [24]. However, our study is based on a much higher number of samples and, therefore, provides much more detail for prioritizing areas for in situ conservation and germplasm collection.

At the allele level, our analysis confirms that, within our study area, the highest allelic richness as well as the highest number of locally common alleles are found in the area of southern Ecuador and northern Peru, i.e. the putative centre of origin of cherimoya. Northern Peru, and more specifically the Cajamarca Department, shows the highest levels of genetic diversity.

The highest values of the fixation index, which is an indication of inbreeding, were found in Ecuador. Inbreeding may occur because of reduction and fragmentation of natural stands and cultivated areas, increasing the risk of allele loss, which eventually leads to genetic erosion [50]. Our results do not allow us to determine how much genetic erosion has taken place in Ecuador in comparison to Peru and Bolivia, but the high inbreeding values in Ecuador could explain why currently allelic richness is lower in this country than in northern Peru.

At the population level, significant differences can be observed between the cherimoya germplasm present in the area with highest diversity (where genotypes belonging to cluster A are predominant) and genotypes found in areas with lower diversity, i.e. in southern Peru and Bolivia (represented by cluster B). Cluster A seems likely to represent material that is genetically closer to the "wild" cherimoya type. No natural cherimoya stands have been observed in Bolivia, and this probably explains why no genotypes pertaining to cluster A have been recorded there. Cluster B probably corresponds to a genepool that is genetically different from most of the wild or semi-domesticated cherimoya found in northern Peru and Ecuador and that could have formed the basis for Bolivian cherimoya cultivation. Looking at the areas of high cluster B dominance, Bolivian germplasm probably originates from southern Peru.

Although most early chroniclers and scientists proposed southern Ecuador and northern Peru to be cherimoya's centre of origin [12], [15], [16], [51], the possibility of that area being a secondary centre of origin cannot be discarded. A diversity study similar to the one described in this study, but including cherimoya genotypes from Central America and Mexico, would shed light on the genetic variation across the complete pre-Columbian distribution range of cherimoya and provide additional clues on the primary centre of origin and diversification of this species.

Ex situ and in situ conservation of cherimoya genetic resources in the Andean region

Most alleles identified in our study are represented in one or more of the existing ex situ collections in Ecuador, Peru and Spain. The results obtained suggest that the highest priority for further collection should be the Azuay Province in Ecuador, since cherimoya stands in this area harbor most alleles not yet included in genebanks. It is also one of the areas with the highest risk of allele loss because of the high observed levels of inbreeding, compared to other parts of the study area. An additional priority area for germplasm collection is the transition zone from the Andes to the Amazon in Peru (in the higher elevation areas of the Departments of San Martin and Amazonas), which was not sampled in this study. According to the distribution model there is a high probability of finding cherimoya stands in this region, which probably is also high in genetic diversity, because it is adjacent to the area with the highest diversity found in this study, i.e. the Cajamarca Department in northern Peru.

A priority for in situ conservation should be the Cajamarca Department, the area with the highest levels of genetic diversity. A second area of priority should be the Loja Province in southern Ecuador, an area with a high number of locally common alleles. Both areas are assigned mostly to cluster A. Since trees pre-dominantly assigned to cluster B have a particular allelic composition in comparison to trees predominantly grouped in cluster A, genotypes of cluster B should also be considered in conservation activities.

The part of Lima Department north of the Peruvian capital, which is assigned mostly to cluster B, could be prioritized for in situ conservation of genotypes from this cluster. In contrast to the low levels of allelic richness around Lima city in the southern part of the Lima Department, the northern part of this Department contains a fair number of locally common alleles.

The long-term conservation of cherimoya genetic resources is far from guaranteed. As commercial prices for fruits can fluctuate, short-term incentives for farmers to maintain cherimoya as a profitable crop are reduced and a decline in commercial interest may lead to the replacement of cherimoya trees by other crops, increasing the risks of genetic erosion. Around Quito, for example, most of the traditional cherimoya cultivation is being replaced by avocado plantations, which are commercially more attractive (X. Scheldeman, pers. obs.). An increase in commercial prices for cherimoya products will not necessarily promote the conservation of the existing genetic diversity. Indeed, in our study we found low levels of genetic diversity around the Peruvian capital, Lima where the clonally propagated cultivar 'Cumbe' is widely cultivated, because it fetches higher prices in the market.

A promising strategy to enhance in situ conservation on farm is through the promotion of seed or bud-for-grafting exchange between farmers [52]. During the CHERLA project, cherimoya fairs, which facilitate exchange of plant material, were organized in different areas of this study, including the Cajamarca and Piura Departments in Peru, Loja Province in Ecuador and various departments in Bolivia. Seed and bud exchange can also be a way to conserve local races from unfavorable alterations in the local environment due to climate change, by re-distributing them in new areas with suitable climate conditions [53]. Another way to ensure conservation of genetic resources of tree species while their use is stimulated could be the establishment of local clonal seed orchards if and when adequate propagation techniques, to enable the multiplication of clones, are made available as well [1], [54]. This is the case for cherimoya, as demonstrated by the successful clonal propagation of the cultivar 'Cumbe' around the city of Lima.

Ideally, each area targeted for in situ conservation - where the existing cherimoya stands and forest patches can evolve within the local environment - should be backed up by ex situ conservation of germplasm (which currently is the case for cherimoya), and be monitored periodically to assess the dynamics in diversity use and risks of genetic erosion. Ex situ collections of fruit tree species often consist of living trees, such as the cherimoya collections. This allows conservation of superior combination of alleles that can be propagated vegetatively through grafting. Additional reasons include the following: (1) many tropical and subtropical trees (including cherimoya) have seeds with recalcitrant or intermediate behavior, which cannot be stored for long-term conservation; and (2) pollen, fruits and seeds can be collected continuously for characterization, evaluation and genetic improvement once trees have reached the reproductive stage. Nevertheless, the high costs for research institutions to maintain field genebanks of woody perennial species, can be a reason to minimize ex situ collections and focus especially on in situ conservation [55]. In that case, it is important to screen the existing accessions through morphological, biochemical and/or molecular characterization to maximize the conservation of genetic diversity and potentially interesting functional attributes in a reduced collection [6]. This approach has already successfully been used in the cherimoya collection la Mayora, Malaga, Spain [27]. Ex situ conservation may particularly be important for areas that suffer from inbreeding -an indicator for high rates of allelic loss and genetic erosion- such as central Ecuador in the case of cherimoya, whereas in situ conservation may be most successful in areas of high diversity where still low rates of inbreeding are observed such as in the cherimoya stands from northern Peru.

Use of GIS and molecular markers to enhance conservation and use of plant genetic resources

Despite the advances in new computational applications and the use of molecular tools, spatial analyses are still underutilized in efforts to conserve plant diversity [56]. With respect to targeting collection sites and prioritizing the conservation of plant genetic resources, spatial analyses of diversity have been carried out mainly at the species level for crop genepools (e.g. [57]–[59]). Only a few studies have mapped intraspecific diversity to enhance the conservation of genetic resources of specific crops and trees (e.g. [36], [41]). Kiambi et al. [41] grouped samples using a grid to compare diversity between geographic areas of similar size, whereas Lowe et al. [36] applied re-sampling to enable the calculation of diversity estimates with high degrees of confidence. However, these studies were carried out with fewer than 100 individuals per species, which limits the type of spatial analysis that can be carried out over the geographic distribution range of species. Our analysis combines both techniques on a large dataset (1504 trees), which can be conceptualized as a continuous distribution of plant individuals, in which each individual is connected to its neighboring trees because they share the same seed system, and/or breed with each other. Based on this concept, trees have been sampled in this study following a scattered distribution to calculate, across the Andean distribution range of cherimoya, several diversity estimates important to prioritize areas for conservation, including two recommended parameters: allelic richness [5] and the number of locally common alleles [43]. Since the application of molecular tools is becoming cheaper, intraspecific diversity studies with large datasets will probably be more common in the near future, allowing for studies of this sort on other tree species and annual crops.

The size of the grid cells and width of the circular neighborhood for this type of spatial analysis depends on how many plant individuals have been collected across the landscape, and the minimum number of plant individuals that is considered sufficient to make confident estimates of genetic parameters per grid cell. Application of circular neighborhood provides an effective way to decrease grid cell size, which facilitates detection of spatial patterns in genetic variation across an extensive distribution range. By re-sampling the trees in the landscape, it generates a high number of grid cells with a sufficient number of trees to make confident calculations of genetic parameters per grid cell. It also makes analyses less sensitive to grid origin definition and enables the inclusion of isolated trees in the calculation of the genetic parameters, i.e. together with their closest neighboring trees.

Ideally, the sampling strategy for this type of analysis should be identified based on a pre-defined grid, aiming at measuring the same number trees per grid cell. However, due to logistical constraints and because a species simply may be more abundant in some areas than in others, in practice, sampling will always be sub-optimal to a certain degree. Of all the genetic parameters, allelic richness is most sensitive to uneven sampling and, accordingly, we have corrected sample size by rarefaction [5]. Repeated subsampling of a minimum number of tree individuals per grid cell is another possibility to correct for sampling bias [60]. This technique could also be used to correct other genetic parameters than allelic richness for sampling bias, such as expected heterozygosity, although these are less sensitive to uneven sampling [61].

Given the sampling distribution in our study area and the fact that for the calculation of most genetic parameters, we maintained a minimum of 20 re-sampled trees per grid cell, we defined a cell size of 10 minutes and a circular neighborhood with a diameter of one degree, which enabled us to detect spatial patterns of genetic variation at administrative level one in Ecuador, Peru and Bolivia (provinces and departments). For studies of plant species, in which individuals are sampled in a more clumped distribution compared to our scattered sampling distribution and/or in lower densities across the landscape, larger grid cells and/or a larger width of circular neighborhood could be applied, always assuring a sufficient number of trees per grid cell. The overall resolution of the study will obviously be lower.

Following Frankel et al. [6], we hypothesized that areas with high diversity measured by neutral molecular markers (like our microsatellite loci) have a high probability to contain genetic material that will also show diversity in functional traits, including traits of agronomic interest. Molecular markers are considered an appropriate indicator to quantify patterns and trends in the use and conservation of plant genetic resources [31]. However, while neutral molecular marker surveys are suitable for diversity studies, direct measurement of traits in field trials may be more desirable to evaluate the genetic health and adaptive capacity of tree populations [50]. Nevertheless, molecular marker studies representative of the whole genome provide a less expensive and scientifically sound alternative to assess the genetic resource status of tree species, for which, in comparison to annual crops, field trials are particularly expensive because of the long generation times [62]. Markers of DNA sequences related to phenotypic traits, including expressed sequence tagged (EST) markers and markers in specific genes, could be of interest to include in spatial analysis of patterns and trends in plant genetic resources. More and more are becoming available, especially for important crops where sequencing programs have been performed or will be carried out in the near future. An example in cherimoya is a recently described gene involved in seedlessness in a sister species, Annona squamosa [63]. However these markers are less polymorphic than neutral ones, such as those that have been used in our study, so the use of neutral markers to study spatial patterns of genetic diversity is still necessary.

It is difficult to compare our results with those of Lowe et al. [36] and Kiambi et al. [41] because of the differences in methodology used. To examine molecular marker studies on the same species, minimum standard sets of markers have already been suggested [64]. Standardization of methodologies in studies on different species would improve comparability of results and also would facilitate Meta-analyses, for example to better understand how well genetic diversity of tropical and subtropical tree species is protected on farm and in protected areas.

In our study we only examined spatial patterns of genetic variation without relating them to other spatial attributes. GIS can also be used to link genetic data to available spatial information relevant to conservation of plant genetic resources, for instance to reveal short-term threats such as accessibility and long-term threats such as climate change. With this type of analysis, hotspots of diversity under threat could be identified following Myers et al. [65] but instead of looking at species level, this could be done at the intraspecific level, to ensure the conservation of priority populations of specific crops and useful tree species. Spatial information on the patterns and characteristics of human societies can be used to understand the drivers behind threats. In a study on changes in cassava diversity in the Peruvian Amazon, GIS was used to correlate cassava diversity data with biotic and socio-economic spatial data to identify possible drivers behind diversity and genetic erosion [66]. This can be useful information in the development of adequate policies and measures to promote in situ conservation of plant genetic resources on farms and in natural populations.

Materials and Methods

Sampling and SSR analysis: A total of 1504 cherimoya accessions have been analyzed in this study, 395 from Bolivia, 351 from Ecuador and 758 from Peru. DNA was extracted from young leaves after

[67]. Based on polymorphism, a set of nine SSRs has been selected from those previously developed in cherimoya [26]. A 15 µL of reaction solution containing 16 mM (NH4)2SO4, 67 mM Tris-ClH pH 8.8, 0.01% Tween20, 2 mM MgCl2, 0.1 mM each dNTP, 0.4 µM each primer, 25 ng genomic DNA and 0.5 units of BioTaq™ DNA polymerase (Bioline, London, UK) was used for amplification on an I-cycler (Bio-Rad Laboratories, Hercules, CA, USA) thermocycler using the following temperature profile: an initial step of 1 min at 94°C, 35 cycles of 30 s at 94°C, 30 s at 45°C–55°C and 1 min at 72°C, and a final step of 5 min at 72°C. Forward primers were labeled with a fluorescent dye on the 5′ end. The PCR products were analyzed by capillary electrophoresis in a CEQ™ 8000 capillary DNA analysis system (Beckman Coulter, Fullerton, CA, USA). Samples were denaturalized at 90°C during 120 s, injected at 2.0 kV, 30 s and separated at 6.0 kV during 35 min. Each reaction was repeated twice and the Spanish cultivar Fino de Jete was used as control in each run to ensure size accuracy and to minimize run-to-run variation.

Data cleaning: The coordinates of the respective tree locations were checked in DIVA-GIS (www.diva-gis.org) on erroneous points based on passport data at administrative level one (e.g. departments, provinces) with a buffer of 20 minutes (approx 30 km), and outliers based on climate data derived from the Worldclim data set [68] (two or more of the 19 bioclim variables according the Reverse jackknife method [69]). Based on these analyses, two points were excluded. The cleaned dataset thus included microsatellite data of 1504 georeferenced trees. Taking into account that nine SSR markers were analyzed, this results in a total of 27,072 georeferenced alleles.

Spatial analysis - Circular neighborhood: Grids for all genetic parameters were generated in DIVA-GIS and are based on a grid with a cell size of 10 minutes (which corresponds to approximate 18 km in the study area) applying a circular neighborhood with a diameter of one degree (corresponding to approximate 111 km) constructed in Excel. The circular neighborhood is used to re-sample the allelic composition of a single tree to all surrounding grid cells, in this case, 32 cells with a size of 10 minutes, within a diameter of one degree around its location. In this way, the allelic composition of each sampled tree is representative for the area within the defined buffer zone. Applying the circular neighborhood re-sampling technique resulted in a total dataset of 48,128 trees and 866,304 alleles.

Spatial analysis - α-diversity: After applying circular neighborhood to all trees, genetic parameters were calculated in GenAlEx per 10-minutes grid cell, for all trees present in each cell after re-sampling. Genetic parameters included the average number of alleles per locus (Na), the number of locally common alleles per locus (alleles occurring with a frequency higher than 5% in 25% or less of the grid cells), average expected heterozygosity per locus (He), fixation index (F) and genetic distance (GD) (see [45]). Na and the number of locally common alleles per locus were presented for all grid cells with trees included. Na was corrected by rarefaction to a minimum sample size of 20 trees per cell with the HP-RARE software (see [70]); consequently, this parameter was only calculated for grid cells with 20 or more re-sampled trees. This minimum sample size was also used as a threshold of the number of trees per grid cell to get interpretable results for the parameters He, F and GD. GD, which was used to calculate distance in allelic composition of each cherimoya genotype to the commercial variety 'Cumbe', was calculated in GenAlEx using the GD option for codominant markers (see [71]). Final GD value per grid cell was the average GD for all re-sampled trees present in each cell. The reference tree was the accession 'Cumbe' from the Spanish cherimoya genebank in Malaga.

Spatial analysis - β-diversity: Population structure was defined by running the software Structure (see [46]) on all 1504 samples applying a 10,000 burn-in period, 10,000 Markov chain Monte Carlo (MCMC) repetitions after burn-in, and 20 iterations. Optimal K was selected after [47] by running Structure for K values between one and 10 and defining the final number of clusters where value of ΔK was highest. This was at K = 2, hence a map was developed for these two clusters, which we named respectively A and B. We used the probabilities of each tree belonging to cluster A and B to visualize the clusters on a map. Mapping of probabilities was done based on the average value of all trees per 10-minutes cell for those grid cells with 20 or more re-sampled trees after applying the one-degree circular neighborhood.

Spatial analysis - Ex situ conservation status: The private alleles function in GenAlEx (PAS) was used to identify the alleles exclusively found in trees that were sampled in situ. To visualize patterns in these alleles that are not included in any genebank, a point-to-grid richness analysis, using a 10-minutes grid, was carried out in DIVA-

GIS based on the one-degree circular neighborhood re-sampled tree grid.

Spatial analysis - distribution modeling: To identify how well the sampling covered the Andean distribution range of cherimoya, and thus to identify potential collection gaps, we modeled the distribution (presence only) of cherimoya in the study area using the distribution modeling program Maxent (see [72], [73]). With this technique, potential distribution areas are identified as those areas where similar environmental conditions prevail as those at the sites where the species has already been observed. The data required to identify these areas include species presence points as well as layers of environmental variables covering the study area. Maxent is a species distribution modeling tool for which the applied algorithm has been evaluated as performing very well, in comparison to other ecological niche modeling software [74], [75]. Therefore, it was selected for this study's distribution modeling analysis. The coordinates in the passport data of the sampled trees were used for the presence point input. For environmental layer input, we used the 10-minutes grids of 19 bioclimatic variables (see [76]), derived from the Worldclim dataset [68]. The modeled distribution area was restricted using the 10 percentile training presence threshold, which indicates the probability value at which 10% of the presence points falls outsides the potential area. The modeled distribution was generated in Maxent with 80% of the points (training data) and was cross-validated in DIVA-GIS with 20% of the remaining tree observations (test data). Besides 20% of the presence points, test data included randomly generated points in 0.1× the bounding box of the presence points as a proxy for absence points (5 times the number of presence points). Based on the cross-validation, the Area Under Curve (AUC) and Kappa value were calculated in DIVA-GIS as measures of model performance.

All maps were edited in ArcMap.

ACKNOWLEDGMENTS

We thank Jorge Rojas and his team from PROINPA for DNA extraction and Bernardo Guzmán from PROINPA for field prospection and sampling in Bolivia. We also thank the personnel from INIA for the DNA extraction and field prospection in Peru and from INIAP Ecuador. Doris Chalampuente, Fernando Paredes, Marcelo Tacán, Eddie Zambrano and Edwin Naranjo. Laura Snook and Evert Thomas from Bioversity, and an anonymous reviewer provided useful comments on an early version of the manuscript.

AUTHOR CONTRIBUTIONS

Conceived and designed the experiments: XS JIH WG CT JR MS MAV. Performed the experiments: PE MV JIH. Analyzed the data: MvZ XS. Contributed reagents/materials/analysis tools: PE MAV JIH WG CT JR MS MvZ XS. Wrote the paper: MvZ XS PVD JIH.

REFERENCES

1. Ræbild A, Larsen AS, Jensen JS, Ouedraogo M, De Groote S, et al. (2011) Advances in domestication of indigenous fruit trees in the West African Sahel. New Forests 41: 297–315. View Article PubMed/NCBI Google Scholar
2. Dawson IK, Lengkeek A, Weber JC, Jamnadass R (2009) Managing genetic variation in tropical trees: linking knowledge with action in agroforestry ecosystems for improved conservation and enhanced livelihoods. Biodiversity and Conservation 18: 969–986. View Article PubMed/NCBI Google Scholar
3. Dawson IK, Vinceti B, Weber JC, Neufeldt H, Russell J, et al. (2011) Climate change and tree genetic resource management: maintaining and enhancing the productivity and value of smallholder tropical agroforestry landscapes. A review. Agroforestry Systems 81: 67–78. View Article PubMed/NCBI Google Scholar
4. Palmberg-Lerche C (2008) Thoughts on the conservation of forest biological diversity and forest tree and shrub genetic resources. Journal of Tropical Forest Science 20: 300–312. View Article PubMed/NCBI Google Scholar
5. Petit RJ, El Mousadik A, Pons O (1998) Identifying populations for conservation on the basis of genetic markers. Conservation Biology 12: 844–855. view Article PubMed/NCBI Google Scholar
6. Frankel OH, Brown AHD, Burdon J (1995) The conservation of cultivated plants. The conservation of plant biodiversity. pp. 79–117. Cambridge University Press, UK. First edition.

7. Tanksley SD, McCouch SR (1997) Seed banks and molecular maps: unlocking genetic potential from the wild. Science 227: 1063–1066. View Article PubMed/NCBI Google Scholar
8. FAO (2010) The second report on the state of the world's plant genetic resources for food and agriculture. Rome.
9. FAO (2011) Draft updated global plan of action for the conservation and sustainable utilization of plant genetic resources for food and agriculture. Fifth session of the Intergovernmental Technical Working Group on Plant Genetic Resources for Food and Agriculture, Rome, 27–29 April 2011.
10. Bremer B, Bremer K, Chase MW, Fay MF, Reveal JL, et al. (2009) An update of the Angiosperm Phylogeny Group classification for the orders and families of flowering plants: APG III. Botanical Journal of the Linnean Society 161: 105–121. View Article PubMed/NCBI Google Scholar
11. Escribano P, Viruel MA, Hormaza JI (2007) Molecular analysis of genetic diversity and geographic origin within an ex situ germplasm collection of cherimoya by using SSRs. Journal of the American Society for Horticultural Science 132: 357–367. View Article PubMed/NCBI Google Scholar
12. Popenoe H, King SR, León J, Kalinowski LS, Vietmeyer ND, et al. (1989) Cherimoya. Lost crops of the Incas: Little-known plants of the Andes with promise for worldwide cultivation. pp. 228–239. National Academy Press, Washington, D.C.
13. Van Damme P, Scheldeman X (1999) Promoting cultivation of cherimoya in Latin America. Unasylva 198: 43–47. View Article PubMed/NCBI Google Scholar
14. Lora J, Hormaza JI, Herrero M (2010) The progamic phase of an early-divergent angiosperm, Annona cherimola (Annonaceae). Annals of Botany 105: 221–231. View Article PubMed/NCBI Google Scholar
15. Popenoe W (1921) The native home of the cherimoya. Journal of Heredity 12: 331–336. View Article PubMed/NCBI Google Scholar
16. Guzman VL (1951) Informe del viaje de exploración sobre la cherimoya y otros frutales tropicales. 25 p. Ministerio de Agricultura, Centro Nacional de Investigación y Experimentación Agrícola La Molina, Lima, Peru.
17. Gepts P (2003) Crop domestication as a long-term selection experiment. In: Janick J, editor. Plant breeding reviews 24 Part 2: Long-term Selection: Crops, Animals, and Bacteria. pp. 1–44.

18. Pozorski T, Pozorski S (1997) Cherimoya and guanabana in the archaeological record of Peru. Journal of Ethnobiology 17: 235–248. View Article PubMed/NCBI Google Scholar

19. Scheldeman X, Van Damme P, Ureña Alvarez JV, Romero Motoche JP (2003) Horticultural potential of Andean fruit crops exploring their centre of origin. Acta Horticulturae 598: 97–102. View Article PubMed/NCBI Google Scholar

20. Vanhove W, Van Damme P (2009) Marketing of cherimoya in the Andes for the benefit of the rural poor and as a tool for agrobiodiversity conservation. Acta Horticulturae 806: 497–504. View Article PubMed/ NCBI Google Scholar

21. CHERLA (2008) Inventory of current ex situ germplasm collections. Deliverable 7, Project no. 015100, INCO sixth framework programme.

22. Pascual L, Perfectti F, Gutierrez M, Vargas AM (1993) Characterizing isozymes of Spanish cherimoya cultivars. HortScience 28: 845–847. View Article PubMed/NCBI Google Scholar

23. Perfectti F, Pascual L (1998) Characterization of cherimoya germplasm by isozyme markers. Fruit Varieties Journal 52: 53–62. View Article PubMed/NCBI Google Scholar

24. Perfectti F, Pascual L (2005) Genetic diversity in a worldwide collection of cherimoya cultivars. Genetic Resources and Crop Evolution 52: 959–966. View Article PubMed/NCBI Google Scholar

25. Escribano P, Viruel MA, Hormaza JI (2004) Characterization and cross-species amplification of microsatellite markers in cherimoya (Annona cherimola Mill. Annonaceae). Molecular Ecology Notes 4: 746–748. View Article PubMed/NCBI Google Scholar

26. Escribano P, Viruel MA, Hormaza JI (2008) PERMANENT GENETIC RESOURCES: Development of 52 new polymorphic SSR markers from cherimoya (Annona cherimola Mill.). Transferability to related taxa and selection of a reduced set for DNA fingerprinting and diversity studies. Molecular Ecology Resources 8: 317–321. View Article PubMed/NCBI Google Scholar

27. Escribano P, Viruel MA, Hormaza JI (2008) Comparison of different methods to construct a core germplasm collection in woody perennial species with simple sequence repeat markers. A case study in cherimoya (Annona cherimola, Annonaceae), an underutilised subtropical fruit tree species. Annals of Applied Biology 153: 25–32. View Article PubMed/NCBI Google Scholar

28. Manel S, Schwartz MK, Luikart G, Taberlet P (2003) Landscape genetics: combining landscape ecology and population genetics.

Trends in Ecology and Evolution 18: 189–197. View Article PubMed/NCBI Google Scholar

29. Holderegger R, Buehler D, Gugerli F, Manel S (2010) Landscape genetics of plants. Trends in Plant Science 15: 675–683. View Article PubMed/NCBI Google Scholar
30. Scheldeman X, van Zonneveld M (2010) Training manual on spatial analysis of plant diversity and distribution. Bioversity International, Rome, Italy.
31. Eaton D, Windig J, Hiemstra SJ, van Veller M, Trach NX, et al. (2006) Indicators for livestock and crop biodiversity. Report.2006/05. CGN/DLO Foundation, Wageningen UR, Wageningen.
32. Kozak KH, Graham CH, Wiens JJ (2008) Integrating GIS-based environmental data into evolutionary biology. Trends in Ecology and Evolution 23: 141–148. View Article PubMed/NCBI Google Scholar
33. Degen B, Scholz F (1998) Spatial genetic differentiation among populations of European beech (Fagus sylvatica L.) in western Germany as identified by geostatistical analysis. Forest Genetics 5: 191–199. View Article PubMed/NCBI Google Scholar
34. Hanotte O, Bradley DG, Ochieng JW, Verjee Y, Hill EW, et al. (2002) African pastoralism: genetic imprints of origins and migrations. Science 296: 336–339. View Article PubMed/NCBI Google Scholar
35. Hoffmann MH, Glaß AS, Tomiuk J, Schmuths H, Fritsch RM, et al. (2003) Analysis of molecular data of Arabidopsis thaliana (L.) Heynh. (Brassicaceae) with Geographical Information Systems (GIS). Molecular Ecology 12: 1007–1019. View Article PubMed/NCBI Google Scholar
36. Lowe AJ, Gillies ACM, Wilson J, Dawson IK (2000) Conservation genetics of bush mango from central/west Africa: implications from random amplified polymorphic DNA analysis. Molecular Ecology 9: 831–841. View Article PubMed/NCBI Google Scholar
37. Vigouroux Y, Glaubitz JC, Matsuoka Y, Goodman MM, Sánchez GJ, et al. (2008) Population structure and genetic diversity of New World maize races assessed by DNA microsatellites. American Journal of Botany 95: 1240–1253. View Article PubMed/NCBI Google Scholar
38. McRae BH (2006) Isolation by resistance. Evolution 60: 1551–1561. View Article PubMed/NCBI Google Scholar
39. van Etten J, Hijmans RJ (2010) A geospatial modelling approach integrating archaeobotany and genetics to trace the origin and dispersal of domesticated plants. PLoS ONE 5: e12060. doi:10.1371/journal.pone.0012060. View Article PubMed/NCBI Google Scholar

40. Guarino L, Jarvis A, Hijmans RJ, Maxted N (2002) Geographic Information Systems (GIS) and the conservation and use of plant genetic resources. In: Engels JMM, Ramanatha Rao V, Brown AHD, Jackson MT, editors. Managing plant genetic diversity. pp. 387–404. International Plant Genetic Resources Institute (IPGRI) Rome, Italy. 2002.
41. Kiambi DK, Newbury HJ, Maxted N, Ford-Lloyd BV (2008) Molecular genetic variation in the African wild rice Oryza longistaminata A. Chev. et Roehr. and its association with environmental variables. African Journal of Biotechnology 7: 1446–1460. View Article PubMed/NCBI Google Scholar
42. Jarvis A, Touval JL, Castro Schmitz M, Sotomayor L, Hyman GG (2010) Assessment of threats to ecosystems in South America. Journal for Nature Conservation 18: 180–188. View Article PubMed/NCBI Google Scholar
43. Frankel OH, Brown AHD, Burdon J (1995) The genetic diversity of wild plants. In: Frankel OH, Brown AHD, Burdon J, editors. The conservation of plant biodiversity. pp. 10–38. Cambridge University Press, UK. First edition.
44. Hijmans RJ, Garrett KA, Huamán Z, Zhang DP, Schreuder M, et al. (2000) Assessing the geographic representativeness of genebank collections: the case of Bolivian wild potatoes. Conservation Biology 14: 1755–1765. View Article PubMed/NCBI Google Scholar
45. Peakall R, Smouse PE (2006) GENALEX 6: genetic analysis in Excel. Population genetic software for teaching and research. Molecular Ecology Notes 6: 288–295. View Article PubMed/NCBI Google Scholar
46. Pritchard JK, Stephens M, Donnelly P (2000) Inference of Population Structure Using Multilocus Genotype Data. Genetics 155: 945–959. View Article PubMed/NCBI Google Scholar
47. Evanno G, Regnaut S, Goudet J (2005) Detecting the number of clusters of individuals using the software STRUCTURE: a simulation study. Molecular Ecology 14: 2611–2620. View Article PubMed/NCBI Google Scholar
48. Araújo MB, Pearson RG, Thuiller W, Erhard M (2005) Validation of species-climate impact models under climate change. Global Change Biology 11: 1504–1513. View Article PubMed/NCBI Google Scholar
49. Fielding AH, Bell JF (1997) A review of methods for the assessment of prediction errors in conservation presence/absence models. Environmental Conservation 24: 38–49. View Article PubMed/NCBI Google Scholar

50. Lowe AJ, Boshier D, Ward M, Bacles CFE, Navarro C (2005) Genetic resource impacts of habitat loss and degradation; reconciling empirical evidence and predicted theory for neotropical trees. Heredity 95: 255–273. View Article PubMed/NCBI Google Scholar

51. Bonavia D, Ochoa CM, Tovar SO, Palomino RC (2004) Archaeological evidence of cherimoya (Annona cherimolia Mill.) and guanabana (Annona muricata L.) in ancient Peru. Economic Botany 58: 509–522. View Article PubMed/NCBI Google Scholar

52. Tapia ME (2000) Mountain agrobiodiversity in Peru. Seed fairs, seed banks, and mountain-to-mountain exchange. Mountain Research and Development 20: 220–225. View Article PubMed/NCBI Google Scholar

53. Mercer KL, Perales HR (2010) Evolutionary response of landraces to climate change in centers of crop diversity. Evolutionary Applications 3: 480–493. View Article PubMed/NCBI Google Scholar

54. Cornelius JP, Clement CR, Weber JC, Sotelo-Montes C, van Leeuwen J, et al. (2006) The trade-off between genetic gain and conservation in a participatory improvement programme: the case of peach palm (Bactris gasipaes Kunth). Forest, Trees and Livelihoods 16: 17–34. View Article PubMed/NCBI Google Scholar

55. van Leeuwen J, Lleras Pérez E, Clement CR (2005) Field genebanks may impede instead of promote crop development: Lessons of failed genebanks of "promising" Brazilian palms. Agrociencia 9: 61–66. View Article PubMed/NCBI Google Scholar

56. Escudero A, Iriondo JM, Torres ME (2003) Spatial analysis of genetic diversity as a tool for plant conservation. Biological Conservation 113: 351–365. View Article PubMed/NCBI Google Scholar

57. Hijmans RJ, Spooner DM (2001) Geographic distribution of wild potato species. American Journal of Botany 88: 2101–2112. View Article PubMed/NCBI Google Scholar

58. Jarvis A, Ferguson ME, Williams DE, Guarino L, Jones PG, et al. (2003) Biogeography of wild Arachis: assessing conservation status and setting future priorities. Crop Science 43: 1100–1108. View Article PubMed/NCBI Google Scholar

59. Scheldeman X, Willemen L, Coppens D'eeckenbrugge G, Romeijn-Peeters E, Restrepo MT, et al. (2007) Distribution, diversity and environmental adaptation of highland papayas (Vasconcellea spp.) in tropical and subtropical America. Biodiversity and Conservation 16: 1867–1884. View Article PubMed/NCBI Google Scholar

60. Leberg PL (2002) Estimating allelic richness: Effects of sample size and

bottlenecks. Molecular Ecology 11: 2445–2449. View Article PubMed/NCBI Google Scholar

61. Lowe A, Harris S, Ashton P (2004) Genetic diversity and differentiation. In: Lowe A, Harris S, Ashton P, editors. Ecological genetics: design, analysis, and application. pp. 50–105. Blackwell Publishing, UK. First edition.
62. Rajora OP, Mosseler A (2001) Challenges and opportunities for conservation of forest genetic resources. Euphytica 118: 197–212. View Article PubMed/NCBI Google Scholar
63. Lora J, Hormaza JI, Herrero M, Gasser CS (2011) Seedless fruits and the disruption of a conserved genetic pathway in angiosperm ovule development. Proceedings of the National Academy of Sciences USA 108: 5461–5465. View Article PubMed/NCBI Google Scholar
64. Van Damme V, Gómez-Paniagua H, de Vicente MC (2011) The GCP molecular marker toolkit, an instrument for use in breeding food security crops. Molecular breeding 28: 597–610. View Article PubMed/NCBI Google Scholar
65. Myers N, Mittermeier RA, Mittermeier CG, da Fonseca GAB, Kent J (2000) Biodiversity hotspots for conservation priorities. Nature 403: 853–858. View Article PubMed/NCBI Google Scholar
66. Willemen L, Scheldeman X, Soto Cabellos V, Salazar SR, Guarino L (2007) Spatial patterns of diversity and genetic erosion of traditional cassava (Manihot esculenta Crantz) cultivation in the Peruvian Amazon: An evaluation of socio-economic and environmental indicators. Genetic Resources and Crop Evolution 54: 1599–1612. View Article PubMed/NCBI Google Scholar
67. Viruel MA, Hormaza JI (2004) Development, characterization and variability analysis of microsatellites in lychee (Litchi chinensis Sonn., Sapindaceae). Theoretical and Applied Genetics 108: 896–902. View Article PubMed/NCBI Google Scholar
68. Hijmans RJ, Cameron SE, Parra JL, Jones PG, Jarvis A (2005) Very high resolution interpolated climate surfaces for global land areas. International Journal of Climatology 25: 1965–1978. View Article PubMed/NCBI Google Scholar
69. Chapman AD (2005) Principles and methods of data cleaning – primary species and species-occurrence data, version 1.0. Report for the Global Biodiversity Information Facility, Copenhagen.
70. Kalinowski ST (2005) HP-RARE 1.0: a computer program for performing rarefaction on measures of allelic richness. Molecular Ecology Notes 5: 187–189. View Article PubMed/NCBI Google Scholar

71. Smouse PE, Peakall R (1999) Spatial autocorrelation analysis of individual multiallele and multilocus genetic structure. Heredity 82: 561–573. View Article PubMed/NCBI Google Scholar
72. Phillips SJ, Anderson RP, Schapire RE (2006) Maximum entropy modeling of species geographic distributions. Ecological Modeling 190: 231–259. View Article PubMed/NCBI Google Scholar
73. Elith J, Phillips SJ, Hastie T, Dudík M, Chee YE, et al. (2011) A statistical explanation of MaxEnt for ecologists. Diversity and Distributions 17: 43–57. View Article PubMed/NCBI Google Scholar
74. Elith J, Graham CH, Anderson RP, Dudík M, Ferrier S, et al. (2006) Novel methods improve prediction of species' distributions from occurrence data. Ecography 29: 129–151. View Article PubMed/NCBI Google Scholar
75. Hernandez PA, Graham CH, Master LL, Albert DL (2006) The effect of sample size and species characteristics on performance of different species distribution modeling methods. Ecography 29: 773–785. View Article PubMed/NCBI Google Scholar
76. Busby JR (1991) BIOCLIM a bioclimatic analysis and prediction system. In: Margules CR, Austin MP, editors. Nature Conservation: Cost Effective Biological Surveys and Data Analysis, CSIRO, Canberra. pp. 64–68.

Chapter 3

NONPARAMETRIC METHOD FOR GENOMICS-BASED PREDICTION OF PERFORMANCE OF QUANTITATIVE TRAITS INVOLVING EPISTASIS IN PLANT BREEDING

[1]Xiaochun Sun, [2]Ping Ma, [1]Rita H. Mumm

[1]Department of Crop Sciences and the Illinois Plant Breeding Center, University of Illinois at Urbana-Champaign, Urbana, Illinois, United States of America

[2]Department of Statistics; University of Illinois at Urbana-Champaign, Urbana, Illinois, United States of America

ABSTRACT

Genomic selection (GS) procedures have proven useful in estimating breeding value and predicting phenotype with genome-wide molecular marker information. However, issues of high dimensionality, multicollinearity, and the inability to deal effectively with epistasis can jeopardize accuracy and predictive ability. We, therefore, propose a new nonparametric method, pRKHS, which combines the features of supervised principal component analysis (SPCA) and reproducing kernel Hilbert spaces (RKHS) regression, with versions for traits with no/low epistasis, pRKHS-NE, to high epistasis, pRKHS-E. Instead of assigning a specific relationship to represent the underlying epistasis, the method maps genotype to phenotype in a nonparametric way, thus requiring fewer genetic assumptions. SPCA decreases the number of markers needed for prediction by filtering out low-signal markers with the optimal marker set determined by cross-validation. Principal components are computed from reduced marker matrix (called supervised principal components, SPC) and included in the smoothing spline ANOVA model as independent variables to fit the data. The new method was evaluated in comparison with current popular methods for practicing GS, specifically RR-BLUP, BayesA, BayesB, as well as a newer method by Crossa et al., RKHS-M, using both simulated and real data. Results demonstrate that pRKHS generally delivers greater predictive ability, particularly when epistasis impacts trait expression. Beyond prediction, the new method also facilitates inferences about the extent to which epistasis influences trait expression.

INTRODUCTION

The estimation of breeding values to facilitate choice of parents is a central problem in plant breeding. Furthermore, in terms of evaluating and identifying outstanding progeny, modern genotyping technologies make it possible to predict performance of new lines based on molecular marker or DNA sequence profile.

Fernando and Grossman [1] first demonstrated the utility of molecular marker data to estimate breeding values in livestock species. These were data involving very few markers. Due to increasingly developed genotyping and sequencing technologies, densely spaced genome-wide SNP (single nucleotide polymorphism) data, involving tens or hundreds of thousands of markers, are now available for a number of crops. The genome-wide markers can be used as 'predictors' to achieve high accuracy in estimating breeding values. However, problems like high dimensionality and multicollinearity emerge when the number of predictors is very large and exceeds the number of records. Therefore, statistical methods to effectively address those issues are urgently needed.

Meuwissen et al. [2] proposed a procedure called Genomic Selection (GS), which uses genome-wide markers to estimate breeding values. Through linear regression of phenotypes on genome-wide markers, this method can model high-dimensional predictors. Then, a shrinkage method can be applied to effectively 'shrink' the effect of multicollinearity and to provide stable parameter estimates [3]. Utilizing these techniques, this approach can generate information about genomic regions that may affect the trait of interest. More recently, other shrinkage methods have been developed to estimate breeding values [4], [5]. These methods are primarily based on linear models, which are easy to interpret and able to fit to the data without overfitting. However, the relationship between breeding value and genetic markers is likely to be more complex than a simple linear relationship, particularly when large numbers of SNPs are fitted simultaneously in the model. While epistasis is recognized as an important source of genetic variation [6], strong genetic assumptions are needed to statistically decompose epistatic variance in those linear models [7], e.g. additive by additive epistasis. Furthermore, the fact that biological basis of epistasis is not well understood makes accommodation for it in the genetic model even more difficult. To address these issues, model-free or so-called nonparametric methods which side-step linearity and require fewer genetic assumptions have gained more and more attention [8]–[10].

Gianola et al. [10] and Gianola and van Kaam [8] first proposed reproducing kernel Hilbert spaces (RKHS) regression for estimating breeding values with genomic data and capturing epistatic interactions. The fundamental idea of the RKHS methods is to replace the original marker values with nonlinear transformed markers through so-called 'basis functions'. After transformation, a new space of predictors is formed and can be used in regression. In reproducing kernel Hilbert spaces, the basis functions are reproducing kernels, which vary according to different inner products defined in the RKHS. Gianola and van Kaam [8] proposed using a multivariate Gaussian kernel suggested by Mallick et al. [11] as a reproducing kernel. However, the multivariate Gaussian kernel assigns equal weights to all predictors. Consequently, one cannot determine which predictors are important and which not by using RKHS regression. Thus, model selection and model building by eliminating unimportant predictors are impossible in this simple framework. Intrinsically, the method becomes a mere function approximation method rather than a statistical modeling method.

Besides RKHS methods, Bennewitz et al. [12] also explored the use of a kernel method which originated from Nadaraya-Watson kernel regression [13], [14] to estimate breeding values. However, the kernel methods incur substantial bias when applied to high dimensional regression with interactions [15].

To overcome the shortcomings of the aforementioned approaches, in this paper, we introduce a smoothing spline ANOVA method in RKHS [16]–[18]. The distinguished feature of the method is decomposition of the multivariate nonparametric function into main effects and interactions, analogous to classic ANOVA in linear models. This gives rise to straightforward interpretation of each component, which distinguishes it from function approximation in Gianola and van Kaam [8]. Assigning different weights to main effects and interactions by some data-driven approach, one can easily conduct model diagnostics, model selection, and model building, However, with tens of thousands of markers in the model, the fitting of RKHS models is computationally expensive, even infeasible [17]. The resulting algorithm is unstable and error-prone. One solution would be to bring down the dimensionality of predictors through usage of a dimension reduction method such as principal component analysis [19]. Alternatively, prediction accuracy may be increased by filtering out 'noisy' markers, an approach supported by results from Meuwissen et al. [2] demonstrating that GS method BayesB which allows certain markers to have no associations with phenotypes had high predictive ability. For the latter, Macciotta et al. [20] and Schulz-Streeck et al. [21] assigned a p-value to each marker through univariate linear regression and used an empirical threshold to remove markers without strong signal. And Long et al. [22] used two steps called "filter" and "wrapper" to select SNPs. Supervised principal component analysis (SPCA) [23] offers both dimension reduction and background noise reduction and serves to supplement RKHS regression.

In this study, we devise and evaluate a two-step method (pRKHS), combining SPCA and RKHS regression, to estimate breeding values and predict phenotypic performance of lines with or without pedigree relationships. In step one, we preselect genetic markers highly correlated with phenotype, and perform principal component analysis on the reduced marker subset. In step two, we use significant principal components as predictors in a smoothing spline ANOVA model to conduct the RKHS regression. The model is fitted using a penalized least squares method, where goodness-of-fit is measured by the least squares and model complexity is dictated by a penalty. The trade-off between goodness-of-fit and model complexity is controlled by smoothing parameters, which are selected by data-driven generalized cross-validation (GCV). The pRKHS method is developed in two versions: pRKHS-NE, which accounts for only additive effects, and pRKHS-E, which includes additive-by-additive interaction effects as well as additive effects in the model. The pRKHS versions are evaluated for predictive ability in simulated genetic scenarios and confirmed in real life scenarios for utility using actual data from corn and barley. The pRKHS versions are also compared in performance with popular shrinkage methods, specifically RR-BLUP, BayesA, BayesB [2] and the nonparametric method RKHS-M used by Crossa et al. [24].

MATERIALS AND METHODS

Simulation

Mating scheme

The breeding scheme for maize line development outlined by Bernardo and Yu [25] was used in the simulation of a number of plant breeding scenarios. Specifically, two unrelated inbreds were crossed to produce an F1 population, from which N doubled haploid (DH) lines (Cycle 0) were generated and crossed to a common tester. Testcross performance data and genotypes of Cycle 0 lines were used to train the model. Based on Cycle 0 testcross phenotypes, Nsel lines were selected to randomly mate for two generations to produce N Cycle 1 lines. Genotypes of Cycle 1 lines were used to predict testcross phenotypes using fitted model. The marker data were coded as zij = 1, if jth marker locus in ith individual was homozygous for marker allele from parental Inbred 1, zij = -1 if homozygous for marker allele from parental Inbred 2 and 0 if heterozygous. N and Nsel values were set as 144 and 8, respectively, according to Bernardo and Yu [25].

Genome model

The genome model for simulation was constructed according to the published maize ISU–IBM genetic map, with a total of 1788 cM [26], with recombination computed using the Kosambi map function [27]. Markers were evenly spaced on the chromosome at 1 cM intervals. And 100 QTLs were randomly positioned across the genome. Both markers and QTLs were assumed to be bi-allelic. The genotypic value for ith individual was calculated according to [7]

Design element is defined according to the general two-allele model (G2A, [28] as. where pk is the allele frequency of Q in kth QTL. Parameter is kth QTL's additive effect and is the epistatic interaction effect between kth and lth QTL. In this case, indicates additive by additive interaction. Furthermore, was sampled from geometric series [25], [29], where L equals the total number of QTL positioned throughout the genome. The direction of effect was randomly assigned to each QTL, leading to random coupling and repulsion linkages. Epistatic effect was sampled from gamma distribution where δ is the indicator function which compares the random variable x generated from uniform distribution [0,1] with P. The extent of epistasis was specified by assigning the proportion (P) of total epistatic interactions with nonzero effect. Three levels of epistasis were considered: P = 0, 0.1 and 0.5, representing no, low, and high epistasis, respectively. The G2A model [28] was chosen to model QTL due to its orthogonal property, which links the genetic variance partition directly to the genetic effect partition. The genetic variance was therefore calculated from the sample variance of genotypic values. Random nongenetic effects were added to the genotypic values to generate phenotypic values in proportion to the heritability (i.e. four heritability levels were considered: 0.1, 0.2, 0.4 and 0.8).

Real Data

To evaluate the predictive ability under real life scenarios, data reported by Crossa et al. [24] on 284 maize lines genotyped with 1148 and 1135 SNPs and phenotyped for anthesis-silking interval (ASI) and grain yield (GY), respectively, were utilized. In addition, three barley datasets generated from North Dakota State University two-rowed (N2) breeding program, with trial name of 'Expt41_2007_Langdon', 'Expt41_2008_Langdon' and 'Exp41_2009_Langdon' from The Hordeum Toolbox (http://wheat.pw.usda.gov/tht/) were utilized. Only entries with phenotypic observations for both grain yield (GYD) and plant height (PHT) from the same location were used to avoid confounding genotype with environment. Trials of year 2007 and 2008 contained different sets of 96 lines while 2009 trial had 57 lines, for a total of 249 unique lines across the three years. Lines of one year were independent from those of others. There were 2161 SNPs for the 2007 dataset, 2029 SNPs for the 2008 dataset, and 1842 SNPs for the 2009 dataset, among which 1641 markers were shared among three years. After filtering out markers with minor allele frequency (i.e. smaller than 0.05), 1511 SNPs were retained for barley data.

SNPs were bi-allelic and the dummy variable for marker data is defined as zij = 1 for A1A1, zij = 0 for A1A2 and zij = −1 for A2A2. For SNP data from BarleyCAP, genotypes '1:1', '2:2', and '1:2' were considered as A1A1, A2A2, and A1A2, respectively. Although the type of marker data is discrete, it is treated as continuous vector of covariates. Missing markers were imputed by averaging marker scores across all lines of that marker. Two missing phenotypes from 2007 and 2008 data were imputed using k-nearest-neighbor (KNN) algorithm [30].

Statistical Methods

pRKHS-E & pRKHS-NE

Features from SPCA and RKHS regression were combined to develop the new method, pRHKS. First, SPCA was applied to reduce the high dimensionality represented by the markers and to decrease 'noise'. Steps to apply SPCA included:

Computing the regression coefficient for each marker on a single marker basis,

Ranking markers by the absolute value of their regression coefficients and selecting a defined number of the top ranked markers to form a marker subset (MS) with which to construct the reduced data matrix, Performing principal component analysis using the reduced data matrix to generate resulting PCs, referred to as supervised principal components (SPCs).

A series of SPCs i.e. explaining 55%,60%, 65%,70%, 75%,80%, and 85% of the data matrix variance were then considered as independent variables to fit a smoothing spline ANOVA model in reproducing kernel Hilbert spaces [17]. Results led to a determination that ~70% (±10%) was an optimal threshold of PC variation explained by the model to achieve high prediction accuracy.

Two versions of the new method (pRKHS-NE and pRKHS-E) were proposed to account for various levels of epistasis.

- pRKHS-NE: All selected SPCs were included in the model as main effects. No interactions were included.
- pRKHS-E: Main effects and two-way, additive-by-additive interactions were included in the model, specifying the level of epistasis. For example, when epistasis was specified as 0.1, then 10% of the epistasis interaction effects were considered to be nonzero. To prevent high dimensionality, each variable and their pair-wise interactions were tested for significance and selected using non-parametric model diagnostics tools, i.e. cosine value (CoV) [17], which corresponds roughly to F-statistics in a parametric regression model. Main effects with CoV larger than 0.05 were retained in the model. To determine the tolerance for various different levels of epistatic interactions, a series of CoV i.e. 0.3, 0.25 and 0.2 were considered.

The nonparametric model pRKHS is written as.

$$Y_i = \eta(\mathbf{x}_i) + \varepsilon_i,$$

where Yi is the phenotype of ith individual, $\mathbf{x}_i = \left(x_i^{(1)}, x_i^{(2)} \ldots, x_i^{(K)}\right)$ is the vector of k SPCs of $x_i^{(1)}, x_i^{(2)} \ldots, x_i^{(K)}$ ith individual, η is some unknown K-variate function relating SPCs and phenotype, and $\varepsilon_i \sim N\left(0, \sigma^2\right)$ is error term for ith individual. We estimate $\eta(\mathbf{x})$ in a functional space $\mathcal{H}$ using a penalized least squares,

$$\frac{1}{n}\sum_{i=1}^{n}(Y_i-\eta(\mathbf{x}_i))^2+\lambda\|\eta(\mathbf{x})\|_{\mathcal{H}}^2,$$

where the first term measures the goodness of fit, and $\|\eta(\mathbf{x})\|_{\mathcal{H}}^2$ quantifies the smoothness of η, and λ is a smoothing parameter balancing the trade-off between the two conflicting goals. Full details of the derived model are described in Appendix S1.

Comparison Methods

pRKHS-E and pRKHS-NE were compared to three shrinkage methods: RR-BLUP, BayesA and BayesB [2]. The general model was written as $\mathbf{y}=\mathbf{1}\mu+\mathbf{X}\boldsymbol{\beta}+\boldsymbol{\varepsilon}$, where $\mathbf{1}$ is a vector filled with ones, $\mathbf{X}$ is marker data matrix, μ is the fixed grand mean, $\boldsymbol{\beta}$ is the vector of marker effects, and $\boldsymbol{\varepsilon}$ is the vector of random residuals. A Gaussian prior was assigned to $\boldsymbol{\beta}$ and $\boldsymbol{\varepsilon}$, with $\boldsymbol{\beta}\sim N\left(0,\mathbf{I}\sigma_\beta^2\right)$ and $\boldsymbol{\varepsilon}\sim N(0,\mathbf{I}\sigma_e^2)$. RR-BLUP was implemented in a Bayesian framework; it assigns a common variance to all marker effects, whereas BayesA and BayesB assigns different variances to different markers. BayesB was modified in this study to include π, the proportion of markers having no genetic variances, as another parameter in the model and assigned it a uniform [0,1] prior instead of arbitrary setting [31]. Variances σ_e^2 and σ_β^2 were assigned a scaled inverse chi-square distribution with scale$S^2$and degree of freedom ν.

Besides shrinkage methods, pRKHS was also compared to the RKHS-M model mentioned in work by Crossa *et al.* [24]. The model was$\mathbf{y}=1\mu+\mathbf{f}+\boldsymbol{\varepsilon}$, where$\mathbf{f}\sim N\left(0,\mathbf{K}\sigma_f^2\right)$. $\mathbf{K}$was the kernel matrix whose $(i,j)th$ entry equaled $\exp\{-\phi*d_{ij}\}$, where $d_{ij}=(\mathbf{x}_i-\mathbf{x}_j)^T(\mathbf{x}_i-\mathbf{x}_j)$ measured similarity between *i*th and *j*th individuals, where $\mathbf{x}_i$ is a vector of marker scores of *i*th individual and bandwidth parameter ϕ was chosen as $2q_{0.5}^{-1}$, where $q_{0.5}$ was the sample median of d_{ij}. And σ_f^2 was also assigned a scaled inverse chi-square distribution with scale $S_f^{\ 2}$ and degree of freedom ν_f.

Data Analysis

The smoothing spline ANOVA model in pRKHS-E and pRKHS-NE was fitted using the *ssanova*function in "gss" package available in R [32]. Default arguments of *ssanova* function were used,*i.e.* argument "*method*" was set to "*v*" to let smoothness parameter λ be selected by GCV and "*type*" was set to "*cubic*" to use a cubic smoothing spline. RR-BLUP, BayesA, and BayesB were coded using C++, among which $S_\beta^2 = 4.23$, $\nu_\beta = 0.05$ and $S_e^2 = 1$, $\nu_e = 1$. The RKHS method was implemented using the program provided by Crossa *et al.* [24], with $S_f^2 = 4$ and $\nu_f = 1$. Gibbs sampler was implemented with 3 chains and 10,000 iterations for each chain to update conditional posterior distributions. The first 1,000 samples of each chain were discarded as burn-in and later thinned by 10. Convergence was checked by inspection of trace plots and Gelman-Rubin plots of error variance using "coda" package in R [33]. Samples from three chains were combined to estimate posterior means. All analyses were run on an Ubuntu Server with 2.8 GHz CPU and 16 GB memory.

In simulation scenarios, ten-fold CV in Cycle 0 (C0) was used to determine the best MS and CoV for pRKHS-E(NE). Pearson correlation coefficients between estimated breeding value (EBV) and true breeding value (TBV) ($r_{EBV:TBV}$), and between EBV and phenotype (PHE) ($r_{EBV:PHE}$) were calculated and averaged across thirty replicated simulations.

Since TBV will never be observed in real cases, the criterion to select MS and CoV was based on $r_{EBV:PHE}$ rather than $r_{EBV:TBV}$. Given the highest $r_{EBV:PHE}$ in C0, optimum MS and CoV were determined and used to estimate breeding values and phenotypes in Cycle 1 (C1). Across a series of MS (*i.e.* from 500 to all markers), percent of variation and number of influential markers whose loadings >0.8*the maximum loading were extracted from top three SPCs, and number of SPC interactions with CoV being 0.2, 0.25 and 0.3 were also recorded.

In real data applications, predicted maize ASI and GY values were based on five-fold CV since only one set of data was available. With barley, $r_{EBV:PHE}$ was computed to measure predictive ability. Expt41_2007_Langdon barley data were used for ten-fold CV to select the optimum MS and CoV for pRKHS-E(NE), which were further used to predict phenotypes of trial Expt41_2008_Langdon and Expt41_2009_Langdon. Model fitting for all real data was repeated five times and results were reported as a mean of five. The R code for computing pRKHS EBVs and correlations with phenotype using the three barley datasets generated by North Dakota State barley breeding program (Expt41_2007_Langdon, Expt41_2008_Langdon and Expt41_2009_Langdon accessed through The Hordeum Toolbox is provided in the supplemental materials (R Code S1).

Results

Simulation Results

Twelve different scenarios were considered in this study to facilitate comparison of methods given various levels of heritability and epistasis (Tables 1, 2, and 3). Pearson correlation coefficients for EBV: TBV ($r_{EBV:TBV}$) and for EBV: PHE ($r_{EBV:PHE}$) were calculated (Tables 1, 2and 3). Both ten-fold CV in C0 and prediction in C1 were used to assess the predictability of the statistical methods.

Heritability	C0/C1	Methods	$r_{EBV:TBV}$ ± SE	$r_{EBV:PHE}$ ± SE
h^2 = 0.1	C0	RR-BLUP	0.474±0.015	0.174±0.019
	C0	BayesA	0.451±0.016	0.170±0.023
	C0	BayesB	0.475±0.015	0.180±0.020
	C0	RKHS-M	0.350±0.021	0.103±0.018
	C0	pRKHS-E	0.422±0.019	0.189±0.005
	C0	pRKHS-NE	0.480±0.016	0.192±0.013
	C1	RR-BLUP	0.329±0.017	0.127±0.018
	C1	BayesA	0.307±0.020	0.124±0.017
	C1	BayesB	0.338±0.017	0.134±0.018
	C1	RKHS-M	0.252±0.014	0.066±0.016
	C1	pRKHS-E	0.342±0.023	0.121±0.026
	C1	pRKHS-NE	0.382±0.016	0.155±0.019
h^2 = 0.2	C0	RR-BLUP	0.572±0.019	0.235±0.018
	C0	BayesA	0.568±0.015	0.230±0.014
	C0	BayesB	0.582±0.018	0.244±0.010
	C0	RKHS-M	0.442±0.012	0.179±0.018
	C0	pRKHS-E	0.494±0.011	0.248±0.013
	C0	pRKHS-NE	0.599±0.018	0.254±0.010
	C1	RR-BLUP	0.470±0.019	0.289±0.015
	C1	BayesA	0.431±0.010	0.265±0.011
	C1	BayesB	0.479±0.020	0.298±0.013
	C1	RKHS-M	0.363±0.018	0.235±0.019
	C1	pRKHS-E	0.341±0.018	0.180±0.011
	C1	pRKHS-NE	0.450±0.019	0.257±0.015
h^2 = 0.4	C0	RR-BLUP	0.785±0.014	0.421±0.018
	C0	BayesA	0.697±0.017	0.354±0.016
	C0	BayesB	0.799±0.016	0.427±0.015
	C0	RKHS-M	0.614±0.017	0.352±0.012
	C0	pRKHS-E	0.756±0.017	0.395±0.011
	C0	pRKHS-NE	0.789±0.020	0.388±0.014
	C1	RR-BLUP	0.614±0.013	0.425±0.011
	C1	BayesA	0.529±0.013	0.361±0.015
	C1	BayesB	0.622±0.013	0.433±0.020
	C1	RKHS-M	0.535±0.022	0.384±0.023
	C1	pRKHS-E	0.513±0.016	0.381±0.016
	C1	pRKHS-NE	0.574±0.018	0.402±0.017
h^2 = 0.8	C0	RR-BLUP	0.827±0.009	0.729±0.006
	C0	BayesA	0.763±0.012	0.673±0.004
	C0	BayesB	0.831±0.009	0.735±0.008
	C0	RKHS-M	0.768±0.011	0.698±0.009
	C0	pRKHS-E	0.678±0.016	0.686±0.012
	C0	pRKHS-NE	0.815±0.014	0.675±0.012
	C1	RR-BLUP	0.744±0.012	0.674±0.014
	C1	BayesA	0.664±0.021	0.601±0.022
	C1	BayesB	0.752±0.011	0.682±0.013
	C1	RKHS-M	0.675±0.010	0.620±0.010
	C1	pRKHS-E	0.633±0.008	0.571±0.011
	C1	pRKHS-NE	0.734±0.010	0.613±0.009

Average correlations ± SE were obtained from thirty replications of each simulation.

doi:10.1371/journal.pone.0050604.t001

Table 1. For scenarios with no epistasis, Pearson correlation coefficients between estimated breeding value and true breeding value (rEBV:TBV) or phenotype (rEBV:PHE) obtained through ten-fold cross-validation with Cycle 0 (C0) and prediction of Cycle 1(C1), implemented for simulated traits with heritability of 0.1, 0.2, 0.4, 0.8, via the various statistical methods.

doi:10.1371/journal.pone.0050604.t001

Heritability	C0/C1	Methods	$r_{EBV:TBV}$ ± SE	$r_{EBV:PHE}$ ± SE
h^2 = 0.1	C0	RR-BLUP	0.418±0.015	0.144±0.009
	C0	BayesA	0.402±0.015	0.134±0.008
	C0	BayesB	0.421±0.014	0.143±0.009
	C0	RKHS-M	0.257±0.012	0.089±0.008
	C0	pRKHS-E	0.433±0.012	0.169±0.018
	C0	pRKHS-NE	0.419±0.015	0.142±0.015
	C1	RR-BLUP	0.369±0.017	0.164±0.010
	C1	BayesA	0.340±0.019	0.153±0.011
	C1	BayesB	0.367±0.018	0.163±0.010
	C1	RKHS-M	0.258±0.018	0.100±0.008
	C1	pRKHS-E	0.394±0.021	0.168±0.005
	C1	pRKHS-NE	0.358±0.017	0.159±0.006
h^2 = 0.2	C0	RR-BLUP	0.535±0.011	0.228±0.019
	C0	BayesA	0.518±0.008	0.234±0.016
	C0	BayesB	0.536±0.011	0.235±0.018
	C0	RKHS-M	0.435±0.014	0.186±0.016
	C0	pRKHS-E	0.542±0.010	0.237±0.015
	C0	pRKHS-NE	0.540±0.010	0.245±0.019
	C1	RR-BLUP	0.512±0.015	0.313±0.016
	C1	BayesA	0.479±0.014	0.267±0.014
	C1	BayesB	0.514±0.015	0.315±0.016
	C1	RKHS-M	0.413±0.010	0.234±0.015
	C1	pRKHS-E	0.484±0.014	0.336±0.006
	C1	pRKHS-NE	0.481±0.014	0.326±0.011
h^2 = 0.4	C0	RR-BLUP	0.688±0.007	0.444±0.008
	C0	BayesA	0.632±0.009	0.421±0.003
	C0	BayesB	0.687±0.006	0.438±0.008
	C0	RKHS-M	0.569±0.011	0.358±0.018
	C0	pRKHS-E	0.696±0.009	0.448±0.008
	C0	pRKHS-NE	0.681±0.011	0.434±0.008
	C1	RR-BLUP	0.606±0.017	0.377±0.015
	C1	BayesA	0.535±0.008	0.327±0.010
	C1	BayesB	0.600±0.020	0.372±0.017
	C1	RKHS-M	0.503±0.013	0.320±0.011
	C1	pRKHS-E	0.605±0.021	0.372±0.020
	C1	pRKHS-NE	0.615±0.016	0.384±0.015
h^2 = 0.8	C0	RR-BLUP	0.802±0.001	0.692±0.002
	C0	BayesA	0.734±0.003	0.633±0.006
	C0	BayesB	0.816±0.004	0.699±0.006
	C0	RKHS-M	0.776±0.004	0.698±0.007
	C0	pRKHS-E	0.809±0.012	0.694±0.012
	C0	pRKHS-NE	0.821±0.007	0.701±0.010
	C1	RR-BLUP	0.770±0.013	0.690±0.012
	C1	BayesA	0.710±0.012	0.634±0.011
	C1	BayesB	0.787±0.013	0.705±0.011
	C1	RKHS-M	0.751±0.014	0.689±0.014
	C1	pRKHS-E	0.775±0.013	0.693±0.010
	C1	pRKHS-NE	0.797±0.014	0.712±0.012

Average correlations ± SE were obtained from thirty replications of each simulation.

doi:10.1371/journal.pone.0050604.t002

Table 2. For scenarios with a low level of epistasis (10% of the epistasis interaction effects are nonzero), Pearson correlation coefficients between estimated breeding value and true breeding value (rEBV:TBV) or phenotype (rEBV:PHE) obtained through ten-fold cross-validation with Cycle 0 (C0) and prediction of Cycle 1 (C1), implemented for simulated traits with heritability of 0.1, 0.2, 0.4, 0.8, via the various statistical methods.

doi:10.1371/journal.pone.0050604.t002

Heritability	C0/C1	Methods	$r_{EBV:TBV}$ ± SE	$r_{EBV:PHE}$ ± SE
$h^2 = 0.1$	C0	RR-BLUP	0.372±0.021	0.175±0.022
	C0	BayesA	0.363±0.020	0.158±0.023
	C0	BayesB	0.336±0.016	0.141±0.015
	C0	RKHS-M	0.173±0.018	0.119±0.013
	C0	pRKHS-E	0.382±0.020	0.203±0.020
	C0	pRKHS-NE	0.363±0.018	0.171±0.021
	C1	RR-BLUP	0.309±0.013	0.182±0.011
	C1	BayesA	0.327±0.019	0.192±0.009
	C1	BayesB	0.298±0.019	0.188±0.010
	C1	RKHS-M	0.157±0.015	0.139±0.008
	C1	pRKHS-E	0.328±0.013	0.194±0.012
	C1	pRKHS-NE	0.298±0.010	0.176±0.011
$h^2 = 0.2$	C0	RR-BLUP	0.487±0.022	0.172±0.020
	C0	BayesA	0.444±0.022	0.175±0.017
	C0	BayesB	0.507±0.024	0.184±0.025
	C0	RKHS-M	0.331±0.026	0.192±0.024
	C0	pRKHS-E	0.512±0.030	0.254±0.023
	C0	pRKHS-NE	0.492±0.024	0.230±0.021
	C1	RR-BLUP	0.416±0.020	0.282±0.011
	C1	BayesA	0.408±0.017	0.256±0.010
	C1	BayesB	0.416±0.008	0.299±0.011
	C1	RKHS-M	0.295±0.011	0.214±0.005
	C1	pRKHS-E	0.441±0.018	0.303±0.010
	C1	pRKHS-NE	0.435±0.014	0.286±0.008
$h^2 = 0.4$	C0	RR-BLUP	0.526±0.016	0.263±0.015
	C0	BayesA	0.520±0.015	0.261±0.021
	C0	BayesB	0.557±0.017	0.300±0.019
	C0	RKHS-M	0.427±0.017	0.306±0.021
	C0	pRKHS-E	0.603±0.016	0.347±0.031
	C0	pRKHS-NE	0.551±0.018	0.333±0.023
	C1	RR-BLUP	0.504±0.022	0.311±0.018
	C1	BayesA	0.462±0.017	0.285±0.014
	C1	BayesB	0.511±0.021	0.315±0.017
	C1	RKHS-M	0.347±0.021	0.267±0.014
	C1	pRKHS-E	0.525±0.016	0.390±0.015
	C1	pRKHS-NE	0.463±0.014	0.344±0.014
$h^2 = 0.8$	C0	RR-BLUP	0.680±0.009	0.407±0.007
	C0	BayesA	0.599±0.008	0.324±0.009
	C0	BayesB	0.697±0.011	0.420±0.008
	C0	RKHS-M	0.584±0.011	0.561±0.012
	C0	pRKHS-E	0.706±0.008	0.535±0.001
	C0	pRKHS-NE	0.660±0.009	0.480±0.010
	C1	RR-BLUP	0.612±0.013	0.298±0.029
	C1	BayesA	0.596±0.014	0.283±0.031
	C1	BayesB	0.637±0.023	0.320±0.028
	C1	RKHS-M	0.475±0.022	0.308±0.053
	C1	pRKHS-E	0.638±0.017	0.418±0.046
	C1	pRKHS-NE	0.618±0.020	0.281±0.036

Average correlations ± SE were obtained from thirty replications of simulation.
doi:10.1371/journal.pone.0050604.t003

Table 3. For scenarios with a moderate level of epistasis (50% of the epistasis interaction effects are nonzero), Pearson correlation coefficients between estimated breeding value and true breeding value (rEBV:TBV) or phenotype (rEBV:PHE) obtained through ten-fold cross-validation with Cycle 0 (C0) and prediction of Cycle 1 (C1), implemented for simulated traits with heritability of 0.1, 0.2, 0.4, 0.8, via the various statistical methods.

doi:10.1371/journal.pone.0050604.t003

For scenarios with no epistasis, BayesB generally outperformed other methods in predictive ability (Table 1). BayesB provided the highest correlation between EBV and TBV in five out of eight cases, except in three cases at low heritability levels (h2 = 0.1 and 0.2), where pRKHS-NE outperformed or at least had comparable results with BayesB. The values of rEBV:TBV for pRKHS-NE were consistently higher than those for pRKHS-E across both C0 and C1, in keeping with the scenario of no epistasis. In three cases where pRKHS-NE outperformed BayesB for rEBV:TBV, it also provided higher correlations between EBV and PHE than BayesB. In no instances did RR-BLUP, BayesA, or RKHS-M provide the highest correlations for TBV or PHE.

For scenarios with epistasis at a low level, the pRKHS method outperformed other methods in predictive ability; particularly in predicting PHE, the pRKHS method provided the highest correlation in all eight cases of C0 and C1 (Table 2). The values of pRKHS-E for rEBV:TBV were marginally higher than those for pRKHS-NE in five out of eight cases of C0 and C1. The pRKHS method provided highest values for rEBV:TBV in seven out of eight cases of C0 and C1, with BayesB providing the highest values in C1 case at h2 = 0.2. For the correlation with PHE, pRKHS-E exceeded pRKHS-NE in three out of four heritability scenarios (h2 = 0.1, 0.2, and 0.4) and performed below pRKHS-NE at h2 = 0.8. In no instances did RR-BLUP, BayesA, or RKHS-M provide the highest correlations for either TBV or PHE.

For scenarios with high epistasis, the pRKHS method, particularly pRKHS-E, outperformed other methods in predictive ability (Table 3). In all cases of C0 and C1 across all heritabilities, pRKHS-E provided the highest correlations for both TBV and PHE. Among four C0 cases, the magnitude of the values of rEBV:TBV of BayesB and pRKHS-NE had decreased the most from the corresponding value in low epistasis scenario at heritability of 0.1 and 0.2, respectively. And RKHS-M experienced the most loss of accuracy at heritability of 0.4 and 0.8. In contrast, pRKHS-E showed the least amount of loss in accuracy based on change of rEBV:TBV in all four C0 cases.

The advantage of marker-based selection (MBS) over phenotypic selection (PS) can be quantified by comparing rEBV:TBV in C1 cases to accuracy of PS, defined as the correlation between mid-parent and offspring and measured by taking square root of half of the narrow-sense heritability [34]. For heritabilities of 0.1, 0.2, 0.4 and 0.8, the accuracy of PS was estimated as 0.224, 0.316, 0.447, and 0.632, respectively. For the scenarios with no and low epistasis (Tables 1 and 2), all methods outperformed PS at all four heritabilities, particularly at low heritability (h2 = 0.1 and 0.2). For the scenarios with high epistasis (Table 3), pRKHS-E and BayesB outperformed PS in all cases, whereas pRKHS-NE and RR-BLUP outperformed PS only at low to moderate heritabilities (h2 = 0.1, 0.2 and 0.4). RKHS-M did not outperform PS at any of the heritability levels.

Table 4 exhibits the range of values for the percentage of variation explained by the first three SPCs, the number of markers included in each of these SPCs, and the number of SPC interactions observed when the cosine threshold for selecting SPC interactions was >0.2, >0.25, and >0.3, respectively, given the series of marker subsets (from as low as 500 markers to all, i.e. 1798 markers) used across 12 simulation scenarios. With low marker density, i.e. 500 markers, the first, second and third SPCs explained up to 18.7%, 11.8% and 9.9%, respectively, of the marker variation. Utilizing the full marker data set, the top three SPCs only explained as low as 9.2%, 5.5% and 5.1% of the variation (Table 4). Averaging across all scenarios, the first three SPC accounted for 25.4% of the marker variation and 18 SPCs were needed to explain 70% of the marker variation (Figure 1). The number of influential markers included in the first three SPCs, namely MP1, MP2 and MP3, varied according to the number of markers used in the model. Using all 1798 markers in simulations, the 1st, 2nd and 3rd SPCs included a maximum of 137, 111, and 103 influential markers, respectively; using the smallest subset of markers i.e. 500 markers, the 1st, 2nd, and 3rd SPCs included as few as 61, 52, and 37 influential markers, respectively (Table 4). The number of SPC interactions was determined by the MS and CoV. High MS or low CoV generate a relatively large number of SPC interactions.

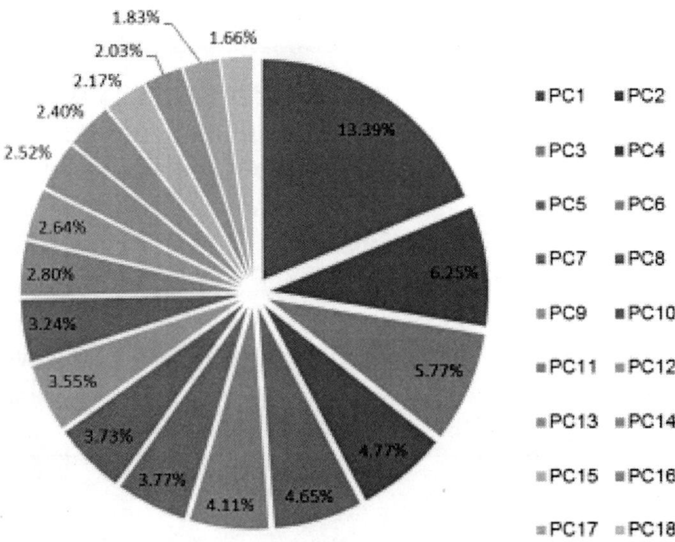

Figure 1. Mean percentage of variation (across the 12 simulation scenarios) explained by the top 18 SPCs with pRHKS, which together explain 70% of the total variation. doi:10.1371/journal.pone.0050604.g001

Scenarios	%P1	%P2	%P3	M_{P1}	M_{P2}	M_{P3}	# of SPC interactions		
							>0.2	>0.25	>0.3
h^2 = 0.1, E = 0	10.4–15.1	5.8–11.1	5.3–9.0	83–127	61–104	43–86	5–12	1–5	0–3
h^2 = 0.2, E = 0	12.2–17.7	5.5–10.8	5.2–8.1	124–136	59–71	56–85	3–11	1–6	0–4
h^2 = 0.4, E = 0	9.3–14.9	6.8–11.7	5.9–9.9	67–111	59–90	56–96	4–16	1–6	0–3
h^2 = 0.8, E = 0	10.4–15.3	5.8–11.0	5.3–9.1	76–124	53–89	48–87	5–20	1–7	0–1
h^2 = 0.1, E = 0.1	11.1–17.7	6.1–9.7	5.4–8.4	105–130	55–98	50–92	4–16	1–5	0–3
h^2 = 0.2, E = 0.1	11.9–16.5	5.6–11.8	5.1–8.2	110–125	66–85	43–88	4–12	2–7	1–4
h^2 = 0.4, E = 0.1	9.2–13.7	6.0–10.4	5.8–9.4	61–122	62–111	53–102	5–18	1–6	1–5
h^2 = 0.8, E = 0.1	11.2–13.0	5.6–10.6	5.1–9.5	69–118	54–77	44–94	6–20	2–8	1–5
h^2 = 0.1, E = 0.5	10.5–14.3	5.7–9.8	5.1–7.8	75–120	57–86	48–103	5–18	2–7	1–2
h^2 = 0.2, E = 0.5	12.0–18.7	6.4–10.2	5.7–7.0	131–137	54–95	37–71	3–17	2–7	1–4
h^2 = 0.4, E = 0.5	12.1–18.5	5.5–11.7	5.0–7.2	122–129	83–99	41–74	5–18	3–7	1–5
h^2 = 0.8, E = 0.5	11.2–18.3	5.8–10.5	5.1–8.6	76–126	52–107	45–96	6–21	2–9	2–5

Values reflect the lows and highs obtained using various marker subsets (from 500 markers to all markers). Note that larger cosine values are equivalent to smaller p-values.

doi:10.1371/journal.pone.0050604.t004

Table 4. For each scenario with pRKHS, the percent of the total variation explained by top three SPCs (%P1, %P2 and %P3), the number of influential markers (MP1, MP2 and MP3) included in the respective SPCs, and number of SPC interactions at three given cosine thresholds.

doi:10.1371/journal.pone.0050604.t004

Results with Field Performance Data

In addition to the simulations, the predictive ability of each method was evaluated using maize and barley data reported by CIMMYT [24] and BarleyCAP, respectively. With the maize dataset, five-fold CV was implemented with the traits ASI and GY to evaluate predictive ability and compare methods. For maize ASI, the RKHS-M method produced the highest correlation of 0.554, followed closely by pRKHS-NE and pRKHS-E (Table 5). For maize GY, the RHKS-M method produced the highest correlation of 0.447, followed closely by pRHKS-NE, RR-BLUP, pRHKS-E, and BayesB.

Trait	CV	Methods	Marker Number	%PC	Correlation
ASI	CV	RR-BLUP			0.495
	CV	BayesA			0.388
	CV	BayesB			0.495
	CV	RKHS-M			0.554
	CV	pRKHS-E	700	70%	0.520
	CV	pRKHS-NE	600	65%	0.526
GY	CV	RR-BLUP			0.423
	CV	BayesA			0.392
	CV	BayesB			0.421
	CV	RKHS-M			0.447
	CV	pRKHS-E	1000	75%	0.422
	CV	pRKHS-NE	900	65%	0.425

The optimal number of markers contributing to phenotypic variation and percent of variations explained by the included SPCs were shown for pRKHS methods; results were averaged across five repeated fittings. Optimal cosine value was 0.3 for pRKHS-E across all datasets.
doi:10.1371/journal.pone.0050604.t005

Table 5. Applying pRKHS to real life scenarios, Pearson correlation coefficients between estimated breeding value (EBV) and phenotype obtained from five-fold cross-validation (CV) implemented for maize anthesis-silking interval (ASI) and grain yield (GY) for each of the 6 statistical methods.

doi:10.1371/journal.pone.0050604.t005

The pRKHS method outperformed RR-BLUP, BayesA, and BayesB with ASI and had comparable performance in GY with RR-BLUP and BayesB, with pRKHS-E and pRKHS-NE performing nearly identically. With regard to pRKHS-E and pRKHS-NE, the optimal numbers of markers contributing to phenotypic variation were 700 and 600, and SPCs explaining 70% and 65% of data matrix variance, respectively, were included in the model in order to achieve high prediction accuracy for ASI trait. More markers were involved in GY, i.e. 1000 and 900 markers for pRKHS-E and pRKHS-NE, respectively, suggesting that more genes and perhaps more epistasis was involved in trait expression (Table 4). Furthermore, the optimum CoV for pRKHS-E was 0.3 for both ASI and GY.

A three-year set of experimental data from BarleyCAP was used to measure the predictive ability given an independent set of breeding lines (Table 6). Phenotypes (i.e. grain yield (GYD) and plant height (PHT)) and genotypes from Year 2007 were used to fit models and evaluate ten-fold CV performance. The fitted models were then used to predict the phenotype of a different set of 96 lines in Year 2008 and 57 lines in Year 2009. For GYD, the pRKHS method substantially outperformed other methods for predicting 2008 and 2009 phenotypes, with pRKHS-E performing better with 2008 predictions and pRKHS-NE performing better with 2009 predictions. The optimal numbers of markers contributing to phenotypic variation were 1500 and 800 and SPCs explaining 70% and 75% of marker variation were included in the model in order to attain high correlation for pRKHS-E and pRKHS-NE, respectively. For PHT, the optimal number of markers was 1000 and SPCs explaining 75% of the variation were included in both pRKHS-E and pRKHS-NE methods. pRKHS-NE generated the highest correlation in 2007 and 2009 data sets whereas pRKHS-E had the highest correlation of EBV and 2008 PHE. Optimal CoV for pRKHS-E was found at 0.3 in both traits.

Traits	Year	Methods	Marker Number	%PC	Correlation
GYD	2007	RR-BLUP			0.449
	2007	BayesA			0.448
	2007	BayesB			0.510
	2007	RKHS-M			0.260
	2007	pRKHS-E	1500	70%	0.438
	2007	pRKHS-NE	800	75%	0.538
	2008	RR-BLUP			0.104
	2008	BayesA			0.073
	2008	BayesB			0.108
	2008	RKHS-M			-0.009
	2008	pRKHS-E	1500	70%	0.295
	2008	pRKHS-NE	800	75%	0.188
	2009	RR-BLUP			0.052
	2009	BayesA			0.085
	2009	BayesB			0.047
	2009	RKHS-M			0.130
	2009	pRKHS-E	1500	70%	-0.081
	2009	pRKHS-NE	800	75%	0.148
PHT	2007	RR-BLUP			0.447
	2007	BayesA			0.446
	2007	BayesB			0.460
	2007	RKHS-M			0.514
	2007	pRKHS-E	1000	75%	0.465
	2007	pRKHS-NE	1000	75%	0.520
	2008	RR-BLUP			-0.015
	2008	BayesA			-0.006
	2008	BayesB			-0.049
	2008	RKHS-M			-0.083
	2008	pRKHS-E	1000	75%	0.084
	2008	pRKHS-NE	1000	75%	0.062
	2009	RR-BLUP			0.076
	2009	BayesA			0.111
	2009	BayesB			0.107
	2009	RKHS-M			0.191
	2009	pRKHS-E	1000	75%	0.203
	2009	pRKHS-NE	1000	75%	0.222

The optimal number of markers contributing to phenotypic variation and percent of variations explained by the included SPCs were shown for pRKHS methods; results were averaged across five repeated fittings. Optimal cosine value was 0.3 for pRKHS-E across all datasets.

doi:10.1371/journal.pone.0050604.t006

Table 6. Applying pRKHS to real life scenarios, Pearson correlation coefficients between estimated breeding value (EBV) and phenotype obtained from ten-fold CV using genotypes and phenotypes of barley lines in year 2007 and prediction based on genotypes of different lines in year 2008 and 2009 implemented for grain yield (GYD) and plant height (PHT) for each of the 6 statistical methods.

doi:10.1371/journal.pone.0050604.t006

Discussion

This study demonstrates the advantages of using nonparametric methods to estimate breeding value and to predict phenotypic performance, especially for traits involving epistatic gene action. The new method is novel because it features a new combination of supervised principal component analysis and reproducing kernel Hilbert spaces, both established statistical methods. The introduction of SPCA complements RKHS by reducing dimensionality and background noise. Two versions of the method were devised to span the range of epistasis involved in trait expression, with pRKHS-E designed to account for low/moderate to high epistasis and pRKHS-NE accommodating circumstances in which no/minimal epistasis exists in the target trait. To evaluate the performance of the pRKHS method, three other shrinkage methods and another nonparametric method RKHS-M were compared. The results obtained from simulation confirmed that in the absence of epistasis, pRKHS-NE performs comparably with BayesB and better than pRKHS-E (Table 1). When epistasis is present, pRKHS-E shows better predictive ability among all other methods (Tables 2, 3). In addition, results with actual data show that the pRKHS method consistently outperforms shrinkage methods and performed comparably to RKHS-M, further confirming the predictive ability of the pRKHS method in real application.

According to selection theory, MBS holds advantage over PS when the genetic correlation (correlation between estimated breeding value and true breeding value) is higher than the correlation of mid-parent and offspring. Given that pRKHS outperformed PS in most of the cases (Tables 1, 2, 3), its potential use to facilitate indirect selection based on marker information alone is highlighted. However, pRKHS-E and pRKHS-NE are not expected to perform equivalently due to different statistical models on which they are based. In the absence of epistasis, overall underperformance of pRKHS-E (Table 1) is mostly attributed to model overfitting. This may be further supported by the results that pRKHS-E had high rEBV:PHE but also the lowest rEBV:TBV among six methods at h2 = 0.8 (Table 1), which suggests estimates from pRKHS-E have higher variance and are more biased in the absence of epistasis. As epistasis was increased in simulation scenarios, rEBV:TBV of pRKHS-E decreased slowly

(Table 2, Table 3), suggesting properly modeling epistasis upholds the advantages of applying MBS.

Note that correlations with pRKHS-E are not overwhelmingly higher compared to pRKHS-NE in low epistasis scenarios (Table 2). The result that pRKHS-E outperforms pRKHS-NE in only five out of eight cases indicates pRKHS-NE may function well when a low level of epistasis impacts trait expression. The above phenomenon may be explained by the fact that the optimal CoV for pRKHS-E was 0.3 in low epistasis scenario, wherein about two to three SPC interactions on average were involved in the model (Table 4). Overall, 18 SPCs were needed to explain around 70% of the variation and these were included as main effects in the pRKHS-E and pRKHS-NE models (Figure 1). Since the principal component score is a linear combination of the weighted marker score, the linear combination of 18 SPC scores may account for a few of the SPC interactions. The above argument is further supported by the observations that fitting model using CoV of 0.2 (i.e. more SPC interactions) causes multicollinearity in some cases. Overall, features of principal component scores may help the additive model pRKHS-NE fit well in the situation of low epistatic interactions.

Cosine threshold value as mentioned in this study is a nonparametric model diagnostic and used as a criterion to select SPC interactions. As the counterpart of F-statistics in a parametric model [17], CoV could theoretically be transformed to a test statistic similar to the F-distribution p-values with some modification [35]. However, the degrees of freedom for F-distribution which are estimated from the trace of the smoothing matrix change every time a new pair of SPC interactions is fitted. Therefore, the consequential p-value is not monotone with the cosine value, indicating the same cosine value could be assigned for different p-values in different model fitting, which is misleading to SPC interaction selection, causing loss or false inclusion of interactions. We did some preliminary experimentation by constructing models using a transformed p-value instead of direct CoV for SPC interaction selection and found low predictive ability (data not shown).

In addition to predictive performance, comparisons between pRKHS and other methods can consider computational load. Several studies [8], [36] have suggested the computational advantages of using nonparametric methods over shrinkage methods. For our models, the cost of the RKHS algorithm is , where n is sample size and q is number of dimensions. With SPCA, q is usually around 18 to 20 (Figure1), indicating computational time of pRKHS will be mainly impacted by sample size instead of marker number. With pRKHS, most of the computational load involves constructing reproducing kernels and the smoothing matrix and estimating smoothing parameter λ. In contrast, the computation load with Bayesian shrinkage methods is linearly related to the number of features since these are Markov Chain Monte Carlo (MCMC) based, with computational time increasing as the number of number of markers increases.

Model performances were influenced by the underlying genetic architecture of the trait of interest. pRKHS plays an important role when trait expression is influenced by epistasis, whereas shrinkage methods may have higher predictive ability when a trait is controlled by strictly additive gene effects. The genetic architecture represented by BayesB assumes a trait is controlled by a few genes with large effects and many genes with small effects. BayesB further allows some of the markers to have zero effect, suggesting a nonuniform distribution of genes contributing to phenotypic variation throughout the genome [2]. Thus, among the three shrinkage methods in this evaluation, BayesB has the most in common with pRKHS with respect to the genetic simulation. Fair approximation of the underlying genome seems to contribute to the good performance of BayesB and pRKHS.

In general, RKHS methods performed better than shrinkage methods. Comparing nonparametric methods, pRKHS performs better than RKHS-M in both simulation and the barley data scenarios (Tables 1, 2, 3, 6) but lower in maize data scenarios (Table 5). These differences in performance might be attributed to two factors: 1) model specifications such as tuning parameter and reproducing kernel, and 2) differences in genetic architecture. The smoothing parameters place different weights to different main effects and interactions, i.e. downplays the effect of unimportant predictors and provides better predictions. The smoothing parameters λ in pRKHS were tuned using the data-driven "GCV" score during model training, while the bandwidth parameter φ in RKHS-M was set to sample median of the squared Euclidean distance as mentioned by Crossa et al. [24]. Meanwhile, we used polynomial kernels to construct kernel matrix while Gaussian kernel was adopted by RKHS-M. To measure the impact of using only one kernel in RKHS-M, kernel averaging model, i.e. K2+K7 developed in de los Campos et al. [36] was also applied on barley data and similar results were obtained (data not shown). It is worth noting that Gianola and van Kaam [8] included parametric mixed effects besides nonparametric function. Such an extension has been also built in the ssanova function [37]. In short, both RKHS-M and pRKHS have their own advantages of prediction, i.e. RKHS-M had higher predictive ability with maize data (Table 5) while pRKHS excel with barley data (Table 6). A combined usage of RKHS-M and pRKHS may outperform the usage of a single method.

The low prediction accuracy observed in "2008" and "2009" as near zero correlation values in Table 6 may be attributed to several reasons. In particular, the three-year barley data are pedigree-independent of each other, suggesting the training and testing datasets are mostly independent. The scenario is different from cross validation within one year (i.e. "2007" correlation values in Table 6), in which case the barley lines are pedigree-related to a certain degree; thus, the training and testing data within a year have more relationship than those across years, leading to higher prediction accuracy.

The predictive ability of pRKHS is highly related to the included SPCs and their interactions (for pRKHS-E). Bair et al. [23] suggested use of the first several SPCs for prediction and later Li et al. [38] applied the first three SPCs in genome-wide association mapping. With our methods, the number of SPCs to include is flexible and depends on the extent of epistasis; it is quantified by selecting the proportion of variation instead of specific numbers. Empirically, we found that with setting a threshold of around 70% as the amount of the variation explained by the model and then utilizing only the SPCs associated with that threshold, an appropriate balance between variance explained and goodness-of-fit was achieved in most of the cases through simulation, and this was confirmed with real data applications (Table 5). However, depending on the crop data and the genetic architecture, the optimal threshold may actually vary by ±10%, indicating a range of 60% to 80% to achieve best predictability. Cross-validation could be used to find the best number of SPCs to include for a specific data.

Optimal marker density for prediction is a topic of great debate. Some studies advocate use of all markers with dense coverage [2], while others found little value in dense coverage of the genome and advocate use of a reduced set of markers for prediction [22], [39], [40]. Ways of selecting markers also vary and can be based on random selection, genetic distance or LD extent, or entropy reduction, for example. In this study, selection of makers was based on the magnitude of the regression coefficient, i.e. the size of the marker effect, and the prediction accuracy is actually increased by discarding certain markers which contribute little to the target trait.

The reduced marker approach used with the new pRKHS method seems to confer some advantages. When no epistasis is present, Bayesian methods perform well with utilization of the full marker information. However, the results that pRKHS-NE had slightly lower prediction accuracy than BayesB (Table 1) suggest a near-similar level of predictive ability may be enabled even if partial marker information is used. With pRKHS methods, a 'preselection' procedure is applied before doing PCA to filter out "non-significant" markers. This increases the probability that the subsequent supervised principal components are in good association with the trait of interest [23]. More importantly, the nonlinearity feature of SPCA which is due to initial marker selection falls into the category of RKHS regression well. Furthermore, PCA serves not only for dimension reduction but also clustering. In simulation, influential markers of each SPC except the first SPC, which contains markers from all ten chromosomes, usually come from one or two linkage groups (chromosomes). Therefore, one SPC is considered to be one or two large haplotypes and the SPC interaction presents the haplotype interactions instead of single marker interaction. It is tedious to evaluate the interaction effect between every pair of SNP markers using a dense marker set; however, pRKHS allows us to narrow down the potential SNP interaction effects by investigating the influential markers of two SPCs which have significant interaction effects with each other. Furthermore, methods using haplotypes have been proved to show higher predictive ability than those only using single marker [41], [42].

Besides prediction, use of pRKHS facilitates inferences about the extent of epistasis involved with a trait of interest. For maize trait ASI with heritability estimated at 0.8 [43], pRKHS-E with optimal cosine value of 0.3 and pRKHS-NE produced comparable results and outperformed RR-BLUP, BayesA, and BayesB that only include additive effects (Table 5), indicating that inclusion of a few pairs of SPC interactions in the model increases prediction. The above results not only correspond to the case of simulated low epistasis scenario with h2 = 0.8 (Table 2) but also are consistent with the conclusions by Buckler et al. [43] who suggested that ASI may involve some low level of epistasis. For GYD and PHT in barley, Xu and Jia [44] concluded that epistasis contributes little to genetic variance for self-pollinated species based on work with a doubled haploid population derived from cultivated parents, although Von Korff et al. [45] later found strong epistatic interactions existed in plant height and yield traits in barley and attributed the reason to use of exotic parents and different statistical approaches. As shown in Table 6, our results align with the low epistasis conclusions from Xu and Jia [44] as pRKHS-E involving a few interactions (cosine value equals 0.3) and pRKHS-NE have comparable predictive ability and both methods are more predictive than others.

Overall, the pRKHS method performs well in estimating breeding value and predicting performance when epistasis explains certain proportion of the phenotypic variation. The rate of genetic gain may be enhanced to a certain degree depending on the underlying epistatic extent. Furthermore, pRKHS can be adapted to different types of genetic architectures, i.e. epistatic extent and linkage disequilibrium, through tuning CoV(representing epistatic strength) and MS (representing the proportion of markers contributing to the target trait), respectively. In cases where no prior knowledge of genetic architecture are known, running methods pertaining to different genetic architectures, such as RR-BLUP (infinitesimal model), BayesB (finite loci model) and pRKHS methods (epistasis model) are recommended. Compared to other methods, pRKHS is not only for prediction purposes but also has the capacity to facilitate inferences about the extent of epistasis involved with a trait of interest, which helps scientists to unravel mysteries about the genetic architecture of complex traits. The new nonparametric methods can be readily extended to account for dominance effects and other semi-parametric methods of dealing with some covariates, e.g. population structure, typically managed in a parametric manner.

ACKNOWLEDGMENTS

We appreciate public access to the genotypic and phenotypic data generated through CIMMYT and the BarleyCAP project supported by USDA NIFA. Thanks to G.R. Johnson for critical review of the manuscript and to anonymous reviewers for their constructive comments which led to improvements in the paper.

AUTHOR CONTRIBUTIONS

Conceived and designed the experiments: XS RHM. Performed the experiments: XS. Analyzed the data: XS PM RHM. Contributed reagents/materials/analysis tools: XS RHM. Wrote the paper: XS PM RHM.

REFERENCES

1. Fernando R, Grossman M (1989) Marker assisted selection using best linear unbiased prediction. Genet Sel Evol 21: 467–477. doi: 10.1111/j.0022-0477.2005.00990.x
2. Meuwissen THE, Hayes BJ, Goddard ME (2001) Prediction of total genetic value using genome-wide dense marker maps. Genetics 157: 1819–1829.
3. Gruber MHJ (1998) Improving efficiency by shrinkage. New York: Marcel Dekker.
4. Xu S (2003) Estimating polygenic effects using markers of the entire genome. Genetics 163: 789–801. doi: 10.1364/OE.18.004469
5. de los Campos G, Naya H, Gianola D, Crossa J, Legarra A, et al. (2009) Predicting quantitative traits with regression models for dense molecular markers and pedigrees. Genetics 182: 375–385. doi: 10.1534/genetics.109.101501
6. Dudley JW, Johnson GR (2009) Epistatic models improve prediction of performance in corn. Crop Sci 49: 1533–1533.
7. Cockerham CC (1954) An extension of the concept of partitioning hereditary variacne for analysis of covariances among relatives when epistasis is present. Genetics 39: 859–882.
8. Gianola D, van Kaam JBCHM (2008) Reproducing kernel hilbert spaces regression methods for genomic assisted prediction of quantitative traits. Genetics 178: 2289–2303. doi: 10.1534/genetics.107.084285

9. Gonzalez-Recio O, Gianola D, Long N, Weigel KA, Rosa GJM, et al. (2008) Nonparametric methods for incorporating genomic information into genetic evaluations: an application to mortality in broilers. Genetics 178: 2305–2313. doi: 10.1534/genetics.107.084293

10. Gianola D, Fernando RL, Stella A (2006) Genomic-assisted prediction of genetic value with semiparametric procedures. Genetics 173: 1761–1776. doi: 10.1534/genetics.105.049510

11. Mallick BK, Ghosh D, Ghosh M (2005) Bayesian classification of tumours by using gene expression data. J Roy Stat Soc Ser B (Statistical Methodology) 67: 219–234.

12. Bennewitz J, Solberg T, Meuwissen T (2009) Genomic breeding value estimation using nonparametric additive regression models. Genet Sel Evol 41: 20. doi: 10.1186/1297-9686-41-20

13. Nadaraya E (1964) On estimating regression. Theor Probab Appl 9: 141–142. doi: 10.1137/1109020

14. Watson G (1964) Smooth regression analysis. Sankhya A 26: 359–372. doi: 10.1137/1109020

15. Fan J (1996) Local polynomial modelling and its applications. Boca Raton, Florida: Chapman & Hall/CRC.

16. Wahba G (1990) Spline models for observational data: SIAM, Philadelphia.

17. Gu C (2002) Smoothing spline ANOVA models. New York: Springer-Verlag.

18. Wang Y (2011) Smoothing Splines: Methods and Applications: Chapman & Hall/CRC.

19. Macciotta NPP, Gaspa G, Steri R, Nicolazzi EL, Dimauro C, et al. (2010) Using eigenvalues as variance priors in the prediction of genomic breeding values by principal component analysis. J Dairy Sci 93: 2765–2774. doi: 10.3168/jds.2009-3029

20. Macciotta N, Gaspa G, Steri R, Pieramati C, Carnier P, et al. (2009) Pre-selection of most significant SNPS for the estimation of genomic breeding values. BMC Proc 3: S14. doi: 10.1186/1753-6561-3-s1-s14

21. Schulz-Streeck T, Ogutu J, Piepho H-P (2011) Pre-selection of markers for genomic selection. BMC Proc 5: S12. doi: 10.1186/1753-6561-5-S3-S12

22. Long N, Gianola D, Rosa GJM, Weigel KA, Avendaño S (2007) Machine learning classification procedure for selecting SNPs in genomic selection: application to early mortality in broilers. J Anim Breed Genet 124: 377–389. doi: 10.1111/j.1439-0388.2007.00694.x

23. Bair E, Hastie T, Paul D, Tibshirani R (2006) Prediction by supervised principal components. J Am Stat Assoc 101: 119–137. doi:

10.1198/016214505000000628

24. Crossa J, de los Campos G, Perez P, Gianola D, Burgueno J, et al. (2010) Prediction of genetic values of quantitative traits in plant breeding using pedigree and molecular markers. Genetics 186: 713–724. doi: 10.1534/genetics.110.118521
25. Bernardo R, Yu J (2007) Prospects for genomewide selection for quantitative traits in maize. Crop Sci 47: 1082–1090. doi: 10.1198/016214505000000628
26. Fu Y, Wen TJ, Ronin YI, Chen HD, Guo L, et al. (2006) Genetic dissection of intermated recombinant inbred lines using a new genetic map of maize. Genetics 174: 1671–1683. doi: 10.1534/genetics.106.060376
27. Kosambi DD (1944) The estimation of map distance from recombination values. Annals of Eugenics 12: 172–175. doi: 10.1111/j.1469-1809.1943.tb02321.x
28. Zeng ZB, Wang T, Zou W (2005) Modeling quantitative trait loci and interpretation of models. Genetics 169: 1711–1725. doi: 10.1534/genetics.104.035857
29. Lande R, Thompson R (1990) Efficiency of marker-assisted selection in the improvement of quantitative traits. Genetics 124: 743–756. doi: 10.1111/j.1469-1809.1943.tb02321.x
30. Hastie T, Tibshirani R, Friedman J (2009) The Elements of Statistical Learning 2nd ed.: Springer.
31. Habier D, Fernando R, Kizilkaya K, Garrick D (2011) Extension of the bayesian alphabet for genomic selection. BMC Bioinformatics 12: 186. doi: 10.1186/1471-2105-12-186
32. Gu C (2011) gss: General Smoothing Splines. R package version 1.1–7. Available: http://CRAN.R-project.org/package=gss. Accessed 2012 Oct 30.
33. Plummer M, Best N, Cowles K, Vines K (2006) coda: output analysis and diagnostics for MCMC. R News 6: 7–11.
34. Falconer DS, Mackay TFC (1996) Introduction to quantitative genetics. Essex, UK: Longman and Company.
35. Ma P, Zhong WX, Liu JS (2009) Identifying differentially expressed genes in time course microarray data. Stat in Biosciences 1: 144–159.
36. de los Campos G, Gianola D, Rosa JMG, Weigel AK, Crossa J (2010) Semi-parametric genomic-enabled prediction of genetic values using reproducing kernel Hilbert spaces methods. Genet Res 92: 295–308.
37. Gu C, Ma P (2005) Optimal smoothing in nonparametric mixed-effect models. Annals of Statistics 33: 1357–1379. doi: 10.1214/009053605000000110

38. Li J, Das K, Fu G, Li R, Wu R (2011) The Bayesian lasso for genome-wide association studies. Bioinformatics 27: 516–523. doi: 10.1093/bioinformatics/btq688

39. Luan T, Woolliams JA, Lien S, Kent M, Svendsen M, et al. (2009) The accuracy of genomic selection in norwegian red cattle assessed by cross validation. Genetics 183: 1119–1126. doi: 10.1534/genetics.109.107391

40. VanRaden P, Van Tassell C, Wiggans G, Sonstegard T, Schnabel R, et al. (2009) Reliability of genomic predictions for North American Holstein bulls. J Dairy Sci 92: 16–24. doi: 10.3168/jds.2008-1514

41. Akey J, Jin L, Xiong M (2001) Haplotypes vs single marker linkage disequilibrium tests: what do we gain? Eur J Hum Genet 9: 291–300. doi: 10.1038/sj.ejhg.5200619

42. Calus MPL, Meuwissen THE, de Roos APW, Veerkamp RF (2008) Accuracy of genomic selection using different methods to define haplotypes. Genetics 178: 553–561. doi: 10.1534/genetics.107.080838

43. Buckler ES, Holland JB, Bradbury PJ, Acharya CB, Brown PJ, et al. (2009) The genetic architecture of maize flowering time. Science 325: 714–718. doi: 10.1126/science.1174276

44. Xu S, Jia Z (2007) Genomewide analysis of epistatic effects for quantitative traits in barley. Genetics 175: 1955–1963. doi: 10.1534/genetics.106.066571

45. Von Korff M, Léon J, Pillen K (2010) Detection of epistatic interactions between exotic alleles introgressed from wild barley (H. vulgare ssp. spontaneum). Theor Appl Genet 121: 1455–1464. doi: 10.1007/s00122-010-1401-y

Chapter 4

BREEDING WITHOUT BREEDING: IS A COMPLETE PEDIGREE NECESSARY FOR EFFICIENT BREEDING?

[1]Yousry A. El-Kassaby, [2]Eduardo P. Cappa, [1]Cherdsak Liewlaksaneeyanawin, [1] Jaroslav Klápšte, [3]Milan Lstiburek

[1]Department of Forest Sciences, Faculty of Forestry, University of British Columbia, Vancouver, British Columbia, Canada

[2]Instituto Nacional de Tecnología Agropecuaria (INTA), Instituto de Recursos Biológicos, Hurlingham, Buenos Aires, Argentina

[3]Department of Dendrology and Forest Tree Breeding, Faculty of Forestry and Wood Sciences, Czech University of Life Sciences Prague, Praha, Czech Republic

ABSTRACT

Complete pedigree information is a prerequisite for modern breeding and the ranking of parents and offspring for selection and deployment decisions. DNA fingerprinting and pedigree reconstruction can substitute for artificial matings, by allowing parentage delineation of naturally produced offspring. Here, we report on the efficacy of a breeding concept called "Breeding without Breeding" (BwB) that circumvents artificial matings, focusing instead on a subset of randomly sampled, maternally known but paternally unknown offspring to delineate their paternal parentage. We then generate the information needed to rank those offspring and their paternal parents, using a combination of complete (full-sib: FS) and incomplete (half-sib: HS) analyses of the constructed pedigrees. Using a random sample of wind-pollinated offspring from 15 females (seed donors), growing in a 41-parent western larch population, BwB is evaluated and compared to two commonly used testing methods that rely on either incomplete (maternal half-sib, open-pollinated: OP) or complete (FS) pedigree designs. BwB produced results superior to those from the incomplete design and virtually identical to those from the complete pedigree methods. The combined use of complete and incomplete pedigree information permitted evaluating all parents, both maternal and paternal, as well as all offspring, a result that could not have been accomplished with either the OP or FS methods alone. We also discuss the optimum experimental setting, in terms of the proportion of fingerprinted offspring, the size of the assembled maternal and paternal half-sib families, the role of external gene flow, and selfing, as well as the number of parents that could be realistically tested with BwB.

INTRODUCTION

Plant breeding, including tree improvement, typically follows the classical recurrent selection scheme, which is characterized by systematic and repetitive cycles of breeding, testing, and selection [1], [2]. These programs deal with multiple populations (e.g., base, breeding, and deployment) and large numbers of parents and offspring, planted over multiple sites and years, and requiring extensive monitoring and maintenance. Selection of elite genotypes for either further breeding and/or inclusion in production populations is commonly performed based on their breeding values, determined from the intra-class correlation among relatives produced from elaborate mating designs [3]. As breeding programs advance, the number of parents' increases and their genealogy overlaps, and mating designs become more elaborate and the time required for their completion become real breeding programs' limiting factors [4]. To alleviate the efforts associated with generating offspring with complete pedigree information, specifically for early generation testing, forest geneticists have adopted simplified protocols, ranging from those not requiring a pedigree (e.g., bulk samples from natural populations known as provenance testing [5] to those with incomplete pedigrees (e.g., open-pollinated [6] or polycross mating [7]). Data analyses with incomplete pedigrees often require invoking and/or accepting un-testable assumptions related to the genetic constitution of the tested families and the numbers of male parents involved in their formation, as well as their proportionate contributions. Since these assumptions are not inordinately realistic in practice, the resulting genetic parameters and their associated inferences are often biased, ultimately leading to various degrees of inaccuracy and inefficiency [8]–[10].

The availability of affordable, highly informative DNA markers, coupled with the development of sophisticated pedigree reconstruction methods, has enhanced their utility in converting incomplete pedigree trials into (effectively) complete trials, thus eliminating the pitfalls associated with the invocation of unfulfilled assumptions [11]. Lambeth et al. [11] initiative of converting the polycross mating design's incomplete pedigree to complete made proper quantitative genetic analyses possible and the method was repeatedly evaluated for several species [12]–[16]. El-Kassaby et al. [13] and El-Kassaby and Lstibůrek [17] capitalized on the restricted maximum likelihood-based "animal model" [18] capability of analysing unbalanced and incomplete pedigree data, along with pedigree reconstruction (tantamount to paternity assignment), to introduce the concept of "Breeding without Breeding (BwB)." The basic idea of BwB is to combine the use of offspring with incomplete pedigree information (an entire open-pollinated test) with a subset of offspring with complete pedigree information, to construct both parental and offspring breeding values, thus incorporating backwards, forwards, and combined selection into an efficient breeding framework [13], [17]. Most of the DNA fingerprinting effort is dedicated to a subset of the offspring from a small number of known maternal parents (seed donors) to generate information about the entire population (maternal and paternal parents, as well as offspring) after reassembling paternal half-sib families from the pedigree reconstruction of the fingerprinted subset. Pedigree reconstruction permits connecting the entire parental population (sampled or not) through their shared offspring thus allowing the implementation of classical quantitative genetics analyses [18].

Here we experimentally demonstrate the utility, the increased precision of genetic parameters estimation, and increased accuracy of predicted breeding values, hence the effectiveness of the "Breeding without Breeding" concept, using open-pollinated offspring from 15 of 41 parents in a western larch (Larix occidentalis Nutt.) "breeding population." We compared the performance of the combined incomplete (half-sib: HS) + complete (full-sib: FS) analysis to that of both the incomplete and complete pedigree designs. Finally, we illustrate the optimum experimental efforts needed for the successful implementation of BwB and discuss the role of factors such as external gene flow, expansion of the test population (i.e., the number of tested parents), and the size of half- or full-sib family needed for accurate genetic parameter determination.

RESULTS

Pedigree Reconstruction/Mating Design Assembly

The partial pedigree reconstruction allowed direct estimation of gene flow, selfing rate, male reproductive success, and the number and/ or size of maternal and paternal half-sib families on the individual as well as the population level (Figure 1). With 95% confidence, 1,419 out of 1,538 (92.3%) fingerprinted offspring were assigned to male parents within the orchard (Figure 1). The remaining 119 paternally unassigned offspring were identified as the product of introgression from an adjacent orchard, suggesting a pollen immigration rate of 7.7%. In addition, a total of 113 individual offspring resulted from selfing (average: 7.4%), ranging from 0.0 to 26.8% among seed donors, reflecting the 15 maternal parents propensity variation to selfing. This variability could be caused by maternal parents' pollen shed and receptivity period synchrony differences.

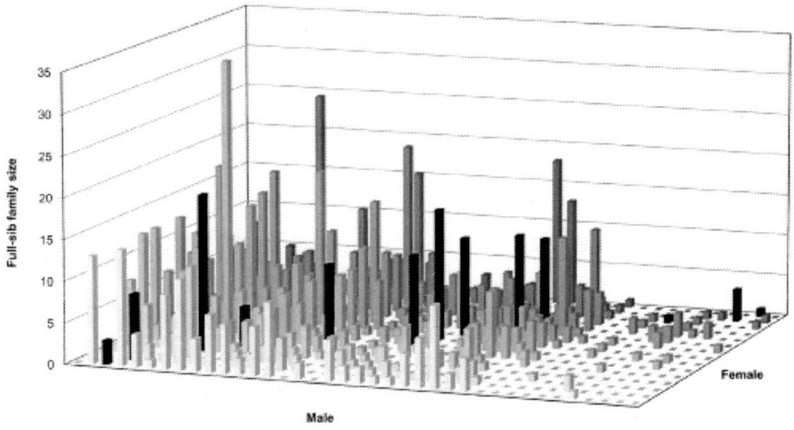

Figure 1. Pedigree reconstruction results showing the formation of full-sib families nested within the maternal and paternal half-sib families (black bars represent selfing).

doi:10.1371/journal.pone.0025737.g001

Pedigree reconstruction resulted in the formation of 349 full-sib families, nested within the 15 maternal and 38 paternal half-sib families, respectively, indicating that three out of the orchard's potential 41 male parents did not participate in pollination, at least of these 15 maternal parents, most likely due to their recent introduction to the seed orchard population (pers. observation). The 15 maternal half-sib families had an average size of 283.9 (range: 222–397) and the 38 paternal half-sib families had an average size of 37.3 (range: 1–193 among the 38 recovered paternal sibships), the latter evidently reflecting male fecundity variation within the orchard. There was an apparently high correlation between the difficult to assess male reproductive investment (male strobili production) and male reproductive success (determined by paternity analysis [19] ($r = 0.87$; $P<0.001$).

The reconstructed pedigree formed a structured mating design, which we used to generate quantitative genetic parameters for the complete pedigree model (FS), and was used in concert with the non-fingerprinted individuals within each of the 15 HS families to form a combined pedigree model, consisting of half- and full-sib families (HS+FS) (see below). A minimum paternal half-sib family size threshold of six individuals was established for inclusion in quantitative genetic analyses. Seven male parents did not meet this threshold, but two were retained, because they were also represented as seed-donors, thus far exceeding the established minimum family size threshold.

Estimation of Quantitative Genetic Parameters

Following the classical individual-tree additive model, three analyses were conducted. The first is for the 15 open-pollinated families (HS) with sample size of N = 5,796 individuals (i.e., incomplete pedigree). The second is also for the same 15 HS families (N = 5,796) but after the inclusion of the male parent for 1,419 individuals (i.e., a combination of half- and full-sib families (HS+FS) and also represents an incomplete pedigree). While the third representing full pedigree (N = 1,419) and was solely based on full-sib families formed by the pedigree reconstruction (FS) (Figure 1; Table 1).

Relative to the combined HS+FS model, the HS model grossly overestimated the additive genetic variance (156.8 vs. 69.3), which more than doubled the height heritability estimate (0.33 vs. 0.14) (Table 1). The precision of the additive genetic variance (80.0 vs. 26.9) and heritability (0.16 vs. 0.05) estimates for these two models produced higher standard error for the HS as compared to the combined HS+FS model (Table 1). Additionally, the inclusion of more genetic information in the combined HS+FS model (i.e., those from FS families) increased the sensitivity of the analysis, as subtle plot-to-plot variation was detected, resulting in a more realistic assessment of the residual error term (Table 1). Parental breeding values' comparisons was limited to only the 15 maternal parents in the HS analysis with their corresponding 15 estimates from the HS+FS analysis and produced non-significant product-moment ($r = 0.44$ (CI: −0.099, 0.775); $p = 0.105$, Figure 2) and rank ($\rho = 0.44$ (CI: −0.099, 0.775); $p = 0.105$) correlations. The corresponding comparison of HS with HS+FS breeding values for the offspring yielded significant product-moment ($r = 0.69$ (CI: 0.672, 0.700); $p = 0.0001$, Figure 3) and rank ($\rho = 0.67$ (CI: 0.656, 0.686); $p = 0.0001$) correlations. Both results clearly demonstrate the reduced utility of the HS model's estimates for forward selection, relative to the results from the HS+FS treatment as indicated by both product-moment and rank correlations. Finally, the average accuracy of predicted breeding values, calculated from the combined HS+FS model was higher for parents (0.81) and offspring (0.55), than their corresponding values (0.56 and 0.45, respectively), calculated from HS model.

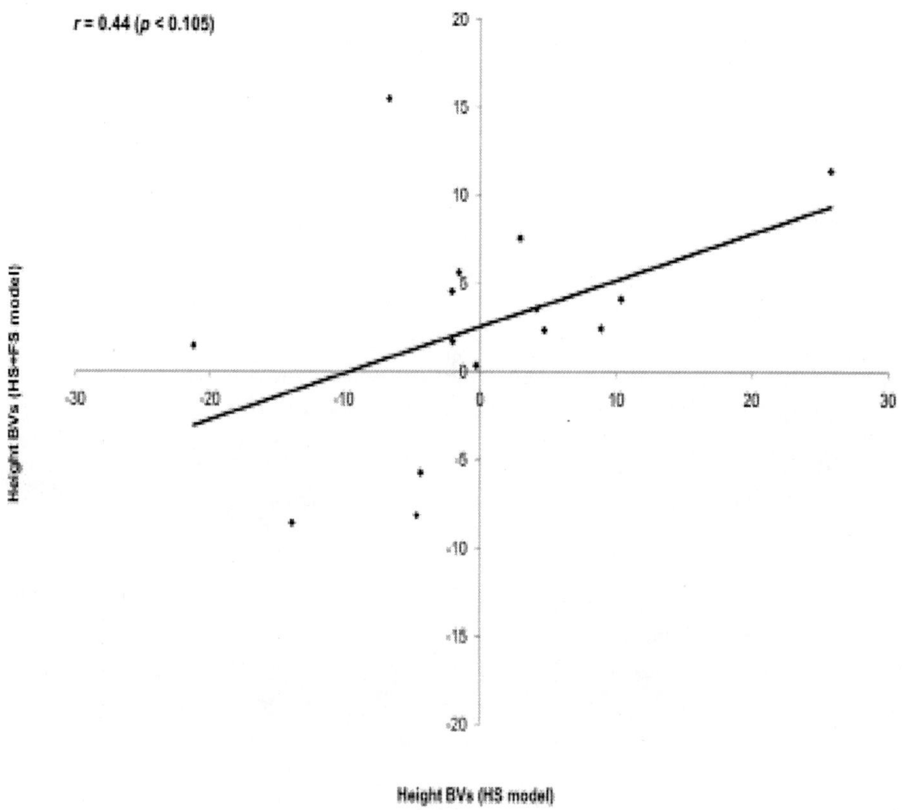

Figure 2. Scatter plot of predicted breeding values for parents from the two incomplete pedigree models (HS and combined FS+HS).

Pearson correlation (r) is in the left corner of the graph.

doi:10.1371/journal.pone.0025737.g002

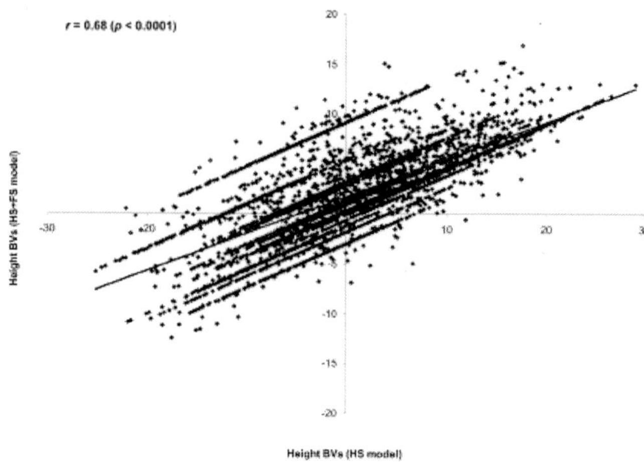

Figure 3. Scatter plot of predicted breeding values for offspring from the two incomplete pedigree models (HS and combined HS+FS).

Pearson correlation (r) is in the left corner of the graph (note the greater extent of variation between the two models).

doi:10.1371/journal.pone.0025737.g003

Source of variation	Variance component		
	Incomplete pedigree		Complete pedigree
	HS	HS+FS	FS
Additive	156.8±80.0	69.3±26.9	55.93±25.42
Plot	48.7±11.2	80.7±17.2	101.95±23.93
Error	266.4±60.5	332.4±20.1	315.99±19.52
Total	471.9	482.5	473.9
h^2_{ns}	0.33±0.16	0.14±0.05	0.12±0.05

doi:10.1371/journal.pone.0025737.t001

Table 1. Forth-year height variance components and narrow sense heritability values (h2ns) and their standard errors for the half-sib (HS), combined half-sib+full-sib (HS+FS) and full-sib (FS) models.

doi:10.1371/journal.pone.0025737.t001

The full (FS) and combined HS+FS pedigree models produced comparable additive and heritability estimates, with similar precision (Table 1). Predictions of parental breeding values extracted from both models were comparable and highly correlated (product-moment (r = 0.96 (CI: 0.928, 0.982); p = 0.0001, Figure 4) and rank (ρ = 0.94 (CI: 0.875, 0.968); p = 0.0001) correlations). The same was true for offspring breeding values (product-moment (r = 0.97 (CI: 0.971, 0.976); p = 0.0001, Figure 5) and rank (ρ = 0.97 (CI: 0.967, 0.973); p = 0.0001) correlations). The results from the combined HS+FS pedigree approach are robust and reliable. Moreover, the average accuracy of breeding values from parents and offspring calculated from the FS model (0.78 and 0.69, respectively) were very similar to those estimated from the combined HS+FS model (0.76 and 0.64, respectively). It is interesting to note that predicted parental breeding values were produced for the entire parental population (i.e., all seed and pollen donors), even when only 15 maternal parents were used and these estimates were based on the entire population (N = 5,796) for the combined HS+FS model as opposed to N = 1, 419 for the FS model.

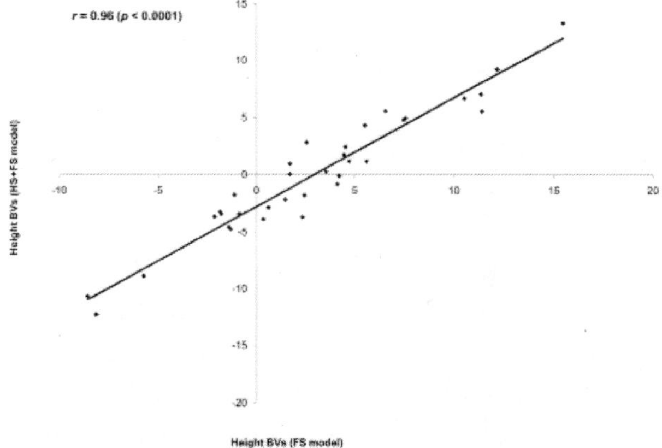

Figure 4. Scatter plot of predicted breeding values for parents from the incomplete (combined HS+FS) and complete (FS) pedigree models.

Pearson correlation (r) is in the left corner of the graph.

doi:10.1371/journal.pone.0025737.g004

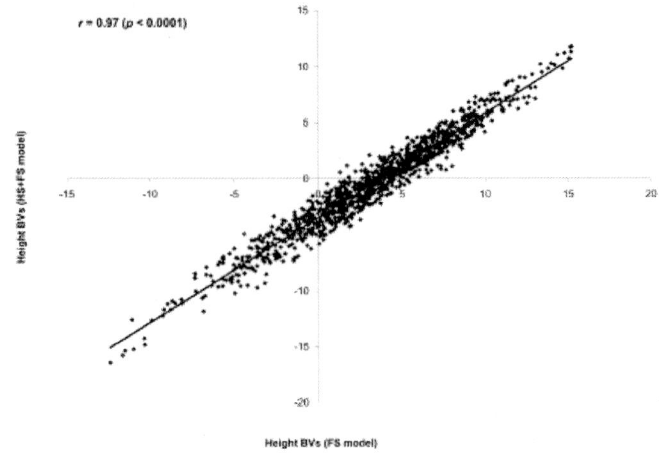

Figure 5. Scatter plot of predicted breeding values for offspring from the incomplete (combined HS+FS) and complete (FS) pedigree models.

Pearson correlation (r) is in the left corner of the graph.

doi:10.1371/journal.pone.0025737.g005

Production Population Selection

We implemented three selection options; namely, forwards, backwards, and combined (combination of backwards and forwards), utilizing either the parental (backwards) and/or offspring (forwards) "Best Linear Unbiased Predictors" (BLUPs) generated from the HS or the combined HS+FS models. The backwards selection option was applied exclusively to the combined HS+FS model as parental breeding values were determined from both maternal and paternal information. The limited number of maternal parents (15 seed donors) precluded the application of the backwards selection option under the HS model; however, maternal breeding values along with offspring was used in the HS combined selection. Additionally, the limited number of maternal parents minimized the response to selection's differences between the forwards and combined selections resulting in somewhat identical results (Figure 6). Without exception and across the range of effective population size tested, the HS model overestimated the response to selection as compared to that from the combined HS+FS model, reflecting the observed additive genetic variance overestimation (Figure 6). For example, compared to the combined HS+FS model, the HS combined selection overestimated the response to selection by a range of 15 and 25% for effective population size of 10 and 40, respectively (Figure 6). The combined HS+FS model's forward and/or combined selections were superior to their backward with response to selection differences ranging between 7 and 12% for effective population size of 10 and 30, respectively (the paternal HS family size restriction of n = 6 limited the effective population size range for backward) (Figure 6). Finally, as expected and for all selection methods and both HS and the combined HS+FS models, the response to selection decreased with increased in effective population size (Figure 6).

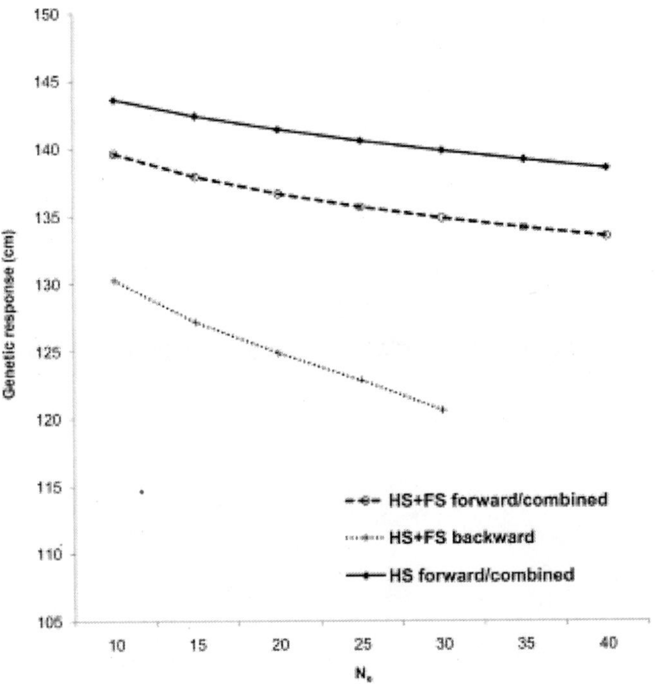

Figure 6. Response to selection comparison between the half-sib (HS) (forward and combined) and combined half- and full-sib (HS+FS) (backward, forward and combined) models assessed across various effective population sizes (10 to 40).

(The small number of tested parents resulted in identical results for forward and combined selection methods under the combined HS+FS and HS scenarios).

doi:10.1371/journal.pone.0025737.g006

Estimating Offspring Optimum Sample Size

Drastic difference in the additive genetic variance magnitude and its standard error was observed with increasing the number of trees with known paternal information (Figure 7). Increasing the number of trees with known fathers (i.e., those from the pedigree reconstruction) to those already with known mothers improved the direct and/or indirect connectedness among parents and thus permitted their unbiased comparison as well as their genetic parameters' estimation. The observed improvement in the additive genetic variance precision leveled after the inclusion of 600 individuals and no substantial fluctuations were observed beyond this point, indicating that a threshold was reached and the inclusion of any additional offspring would not substantially affect the results (Figure 7). Based on the observed trend and in this particular case, it appears that the inclusion of paternal information for 10% of the evaluated offspring population is adequate to create the direct and/or indirect connectedness among parents is sufficient to achieve the available precision.

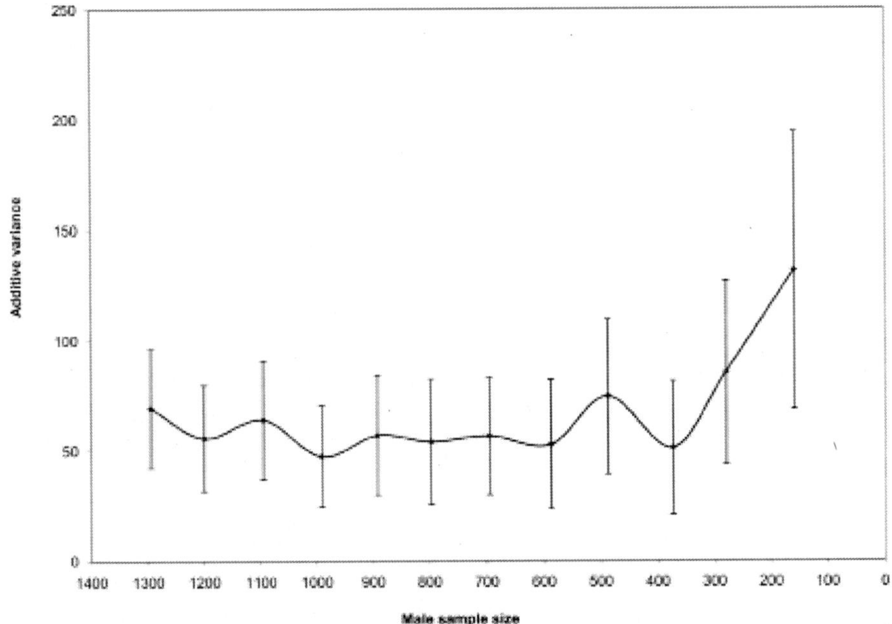

Figure 7. Additive genetic variance estimates as affected by variation in the number of offspring with known male parents.

Vertical lines represent the standard error bars for additive genetic variance estimates.

doi:10.1371/journal.pone.0025737.g007

Discussion

The concept of marker-assisted estimation of quantitative genetic parameters was introduced by Ritland [20], whereby traits' heritabilities and the magnitude and direction of their genetic correlations are derived from regressing pair-wise phenotypic similarity on their corresponding pair-wise genetic relatedness. This concept is appealing, because of its obvious simplicity, in situ nature (i.e., no experiments or mating designs), and most of all its suitability to long-lived organisms such as trees or wildlife that require long-term experiments or extensive field observations. The distribution of relatedness among the studied individuals is assumption-free, thus it is applicable to natural populations where a vast array of genetic relationships can occur [21], [22]. In situations where offspring are derived from random mating among a set of known parents and more specifically when their number is somewhat limited, the no a priori assumption about the expected distribution of genetic relationship becomes inappropriate for a network of full-sibs, half-sibs, and selfs (albeit absence of spatial autocorrelation in relationship coefficients as well as in trait performance in the wild are assumed). It should be stated; however, that the regression approach does not permit the prediction of parents and/or offspring breeding values, thus its application to selection and breeding is somewhat limited.

Conventional tree breeding programs are structured around three main activities: breeding, testing and selection [23]. These activities are long-term endeavours, based on structured pedigree produced from one or a combination of different mating designs [23]; they also require extensive testing in large experimental settings, distributed throughout vast territories [4], and (most important of all), they require sustained organizational and financial commitment. Obviously, simplified breeding schemes that reduce time and cost would be of great value. The generation of complete pedigreed offspring for testing and selection is an obvious target for simplification, fostering incomplete pedigree methods such as open-pollinated family testing [6] and polycross designs [7]. Incomplete pedigree methods, however, are not without their limitations. In particular, open-pollinated testing of the offspring of each maternal parent (seed-donor) assumes they are half-sibs (i.e., sired by different fathers) and non-inbred, and the covariance among half-sib families is assumed to be equal to one-fourth of the additive genetic variance [3]. Both theoretical and empirical studies indicate that this assumption is often violated; as a practical consequence, additive genetic variance is typically overestimated [8]–[10]. The extent of the half-sib assumption violation is expected to be greater if the number of male parents is restricted, as it typically is in a confined breeding population, the usual strategy in breeding arboreta.

To avoid the inaccuracies associated with quantitative genetic parameters assessment from incomplete pedigrees, Lambeth et al. [11] proposed the use of molecular genetic markers for paternity assignment, thus converting the incomplete to a complete pedigree, allowing proper genetic parameters estimation and reliable parental and offspring ranking. The same approach was also introduced to open-pollinated testing by Grattapaglia et al. [12], who reconstructed the complete pedigrees.

El-Kassaby et al. [13] and El-Kassaby and Lstibůrek [17] introduced the concept of "Breeding without Breeding" as a simple, alternative scheme to conventional tree breeding. The method uses: 1) large open-pollinated (i.e., incomplete pedigree) as a primary mean to simplify testing, 2) informative DNA fingerprinting and pedigree reconstruction for a randomly selected subset of the tested individuals to determine their genetic relationship (i.e., complete pedigree) and hence provide adequate bridges between all parents (female and male), 3) the animal model [18] to concurrently analyse the combination of complete (FS: subset) and incomplete (HS: open-pollinated families) pedigree to generate the quantitative genetic parameters needed for selection, and 4) application of an optimization protocol [17] that maximizes the genetic gain at any desired genetic diversity level in a selection scheme. The method capitalizes on the animal model's [18] capabilities of analysing unbalanced and incomplete pedigree to generate the genetic parameters using the "Best Liner Unbiased Prediction" procedure (BLUP [24] needed for parental and offspring evaluation thus facilitating backwards, forwards, or a combined (backwards and forwards) selection in a breeding framework. Therefore, the fundamental difference between the assembled genetic relationship among individuals in the BwB scheme (present study) and those from either the polycross [11], [16] or open-pollinated testing [12]–[16] is that the former does not require complete pedigree for the tested population (a combination of large half-sibs and several smaller full-sibs families) while the latter explicitly stipulates the availability of complete pedigree information for every individual for quantitative genetic parameters estimation.

Quantitative genetic parameters comparison between the two incomplete-pedigree models (i.e., HS and the combined HS+FS) indicated that the HS model over-estimated the additive genetic variance and its surrogate heritability and under-estimated the environmental effects (Table 1). As expected, the genetic relationships (half-sib, and full-sib; Figure 1) within the studied 15 half-sib families should have reduced the average covariance among relatives within the HS model, thus the resulting additive genetic variance is unrealistically inflated. Furthermore, the HS model failed to detect the subtle site heterogeneity present in the experimental site [25], hence the observed under-estimation of the plot effect (Table 1). This is due to the fact that the 15 half-sib families were present in 4 large, 10×10 replications which made it difficult to definitively separate the genetic and environmental effects within experimental units (i.e., plots). In multiple-tree and contiguous plots designs, substantial environmental covariance among family members is confounded with genetic covariance of a given plot [25].

The degree of confounding depends on the size of the plots and the patterns of environment variability. In general, the larger the plot, the more difficult it is to cleanly separate genetic from environmental effects. On the other hand, site heterogeneity was clearly detected after the inclusion of more genetic information in the combined HS+FS model (i.e., those resulting from the pedigree reconstruction of 1,419 individuals which resulted in a better site heterogeneity detection due to their presence across all half-sib families and their respective replications). It is noteworthy to mention that the changes in variance components apportionment over the HS and the combined HS+FS models' sources of variation, collectively affected the resulting heritability estimate (Table 1). While it is only for a subset of the offspring, the inclusion of additional paternal information in the combined HS+FS model permitted covariance among relatives adjustment and hence the observed improvement in the generated parameters, a situation cannot be attained under the HS model (i.e., open-pollinated test).

The discrepancy between the two models is further demonstrated by the low to moderate correlations between either paternal or offspring breeding values (Figs. 2–3) and their different average accuracy of prediction (0.56 vs. 0.81 for parents and 0.45 vs. 0.55 for offspring), highlighting the reduced reliability of the open-pollinated testing for either backwards or forwards selection. Furthermore, the combined HS+FS model allowed predicting the breeding value for the entire parental population (38 vs. 15) as it utilized all offspring information irrespective of parental gender (i.e., as pollen and/or seed donors) while the HS model was restricted only to the maternal population (i.e., seed donors).

The observed differences between the two incomplete pedigree models (HS and the combined HS+FS) support the beneficial role of including the pedigree reconstruction information even though it is only from a subset of the studied population.

The inclusion of additional genetic information allowed the creation of linkages among the 15 half-sib families (known seed donors) with all parents participated in mating (pollen donors), thus increasing the sample size (i.e., higher genetic parameters' precision and breeding values' accuracy) and maximizing the BLUP-method utilization for breeding values prediction (see Ronningen and Van Vleck [24], for detailed explanation). The comparison between the combined HS+FS and full pedigree (FS) models is also needed to illustrate the advantages of partial pedigree inclusion. The full pedigree (FS) model is based on the assembled mating design from the pedigree reconstruction that is based on 1,419 offspring. Variance components and their precision and parental and offspring breeding values comparison between the two models produced similar estimates (Table 1; Figs. 4–5) and accuracies for parents (0.76 vs. 0.78) and offspring (0.64 vs. 0.69) were virtually identical. Heritability estimates are known to be population-specific [3]; however, the two models produced comparable 4-year height heritability estimates (HS+FS: 0.14±0.05; FS: 0.12±0.05), indicating similar magnitude/trajectory. This is not surprising since the two populations share 1,419 individuals in common and the combined HS+FS model included additional 4,258 individuals with known maternal parents. More importantly, the striking similarity between parental and offspring breeding values between the two models are indicative of similar ranking even though different number of individuals and genetic information were used. The observed high correspondence between the suggested combined HS+FS and complete pedigree models highlights the superiority of the proposed BwB [17] indicating that a mixture of incomplete (half-sibs) and complete (full-sibs) pedigree is an efficient approach for acquiring reliable quantitative genetic parameters. The fingerprinting of a subset of the testing population is expected to substantially reduce the cost associated with pedigree reconstruction without any parameters' precision penalties.

The advantage of the combined HS+FS model over the HS and/or FS models is clearly demonstrated at the selection stage (Figure 6). Notwithstanding the overestimation of the additive genetic variance, the HS model is restricted to backward selection from the studied female parents as no BLUP values are generated for their male counterparts (i.e., 15 out of 38). The FS model is better than the HS as it allows the generation of accurate BLUP values for the 38 parents participated in mating as well as their offspring (N = 1,419) which is a subset of the tested population (N = 5,796), thus limiting forward selection to the fingerprinted offspring and thus does not consider any of the non-fingerprinted offspring which represent a substantial part of the tested population (57%). The combined HS+FS model, on the other hand, provides BLUP values for the parents and their offspring, irrespective of their family status, thus increasing the efficiency of forwards selection and improving the precision of backwards selection as well as combined selection. Additionally, the establishment of open-pollinated vs. those based of full pedigree field tests is more simplistic and can be effectively done with reduced efforts and cost.

The large number of parents commonly tested in traditional tree improvement programs requires the use of "efficient" mating designs so manageable number of crosses are made (e.g., disconnected partial diallel [4], [23]). In these mating designs, the parental population is divided to multiple subsets of parents with crosses are often restricted to within parental subsets with minimal or no matings among members of the different subset, thus creating opportunities for genetic sampling (i.e., no opportunity for cross referencing across set). The present study has demonstrated that paternity assignment of wind-pollinated half-sib families from know seed-donors provided adequate linkage across parents, hence we propose the implementation of similar approach concurrently with the selected traditional mating schemes to provide means for cross referencing and the avoidance of genetic sampling.

If BwB is to be considered as a viable option for tree breeding, then several additional questions must to be answered, among them: 1) what is the proportion of the population needed for pedigree reconstruction? 2) What is the minimal HS and/or FS family size required for proper BLUP analysis? 3) What is the role of elevated gene flow or selfing in the breeding population? 4) How many parents can be realistically tested? 5) How are we to expand testing beyond those parents present in the breeding population? The observed changes in the additive genetic variance estimates and their associated precision that accompany changes in the number of genotyped individual with known male parents (i.e., those resulting from the pedigree reconstruction) suggest that the inclusion of approximately 10% of the tested population is adequate to reach stable parameter estimates (Figure 7). The main function of these individuals is to create enough connections between parents, thus permitting direct and/or indirect comparison among the parental population members, a fundamental prerequisite for the BLUP analysis [24]. Increasing the number of offspring with known fathers to those already with known mothers increased the direct and/or indirect connectedness among parents and thus permitted their unbiased comparison as well as the estimated genetic parameters. Rönningen and Van Vleck [24] explicitly stated that a minimum of two offspring between any two males is needed for proper parameters estimation. In the present analysis, we imposed a minimum half-sib family size of six for any parent to be included and the observed correspondence between parents and offspring breeding values between the combined HS+FS and FS models is a reflection of this practice. The number of offspring designated for fingerprinting will also be affected by the degree of gene flow. As gene flow increases, more genotyped individuals will not provide any paternal information for connecting the different parents, but those individuals will remain in the analysis if they are among the maternal parents evaluated. Additionally, as the selection differential between the gene flow's source and the parental breeding population increases, the greater the difference in their offspring performance. A simple offspring phenotypic ranking followed by truncation selection theoretically could eliminate a substantial amount of the inferior offspring [17]. Offspring produced through selfing, while limited, remained in the data analysis through the inclusion of the pedigree information, and thus the estimated genetic parameters should be minimally affected. The rate of selfing among the tested parents is expected to provide an idea of the selfing propensity variation, which may shed some light on the relationship between selfing rate and general combining ability. As the number of parents' increases, the number of informative genetic markers must increase to allow for the exclusion power needed for pedigree reconstruction. The use of paternally inherited markers such as cpDNA could aid in differentiating among males with similar autosomal multilocus genotypes. Increasing the number of marker loci and including paternally inherited markers is expected to increase the experimental efforts; however, the increased efforts should be evaluated in light of the number of parents tested. Finally, increasing the number of tested parents beyond what is present in the breeding population could be accomplished through the use of supplemental-mass-pollination, a technique known to successfully incorporate pollen from specific parents in natural wind pollination of unprotected receptive females [26].

MATERIALS AND METHODS

Plant material

In 2005, wind-pollinated seed samples from 15 unrelated parents were collected from a 41-parent western larch (Larix occidentalis Nutt.) seed orchard. The sampled orchard is one of two genetically distinct (41 and 62 parents) orchards established by British Columbia Ministry of Forests, Lands and Natural Resource Operations to provide genetically improved seed to the Nelson (<1,300 m) and East Kootenay (800–1,500 m) seed production units. These orchards are located near Vernon, B.C., Canada (altitude 480 m, latitude 50°14′N, longitude 119°16′E), in an area devoid of western larch background pollen. The orchards are separated by an 8 m wide road and a row of black cottonwood (Populus trichocarpa Torr. & Gray) trees, acting as a partial pollen barrier. Seed samples and orchard's reproductive survey data were provided by British Columbia Ministry of Forests, Lands and Natural Resource Operations as the orchard is part of a co-operative arrangements among government-private industry-academia. Seed were sown (February, 2006) by individual maternal family in a commercial nursery in growing blocks (80 cavities/block), soil mixes, irrigation, heating, and fertilization regimes similar to those operationally applied for reforestation seedling production. Seed pre-treatment (i.e., pre-chilling to break dormancy) prior to sowing followed International Seed testing Association procedures [27]. At the end of the growing season (September, 2006), seedlings were extracted, by family, and used to establish a common garden trial.

Common garden trial

The trial was established at the University of British Columbia's Research Facility (latitude 49° 15'N, longitude 123° 15'W, elevation 79 m), laid out as a randomized complete design with four replications. Each replication consisted of 10×10 square plots at a spacing of 0.3×0.3 m (100 seedlings/family). At the end of the third field growing season (fall of 2009, 4 years from germination), total seedling heights (HT in cm) were measured on all surviving trees (5,306). The trial was watered and weeded when needed, and survival was 88% at the time of height measurement.

DNA fingerprinting and paternity assignment

The two orchards (studied and neighbouring, with their 41 and 62 parents, respectively) represent the possible paternal parents for a randomly selected 1,538 offspring that were genotyped with 16 microsatellite (SSR) markers. The SSR markers used were: 1) seven developed for Larix occidentalis [28], 2) two developed for L. lyalli [29], with one primer (UAKLly13) amplifying two loci (UAKLly13-1 and UAKLly13-2) in L. occidentalis, and [3] seven developed for L. kaempferi [30] (Table S1). Touchdown PCR was performed according to the protocol used by Isoda and Watanabe [30]: 94°C for 1 min followed by 10 cycles/30 s at 94°C, 30 s/63°−53°C (−1°C at each cycle) for 1 min, followed by 25 cycles of 30 s/94°C, 30 s/53°C and 1 min/72°C followed by 10 min/72°C. The CERVUS program ver. 3.0.3 was used to estimate null allele frequencies in the studied orchard's parental population [31]–[32], as null alleles introduce errors in parentage analysis by leading to high frequencies of false parentage exclusions [33]. PCR multiplexing was developed for four sets of loci sharing the same annealing temperature: 1) UBCLXdi-16, UBCLX1-10, and UBCLXtet-21, 2) UBCLXtet_2-12, UAKLly10, and UAKLly13, 3) bcLK33, bcLK66, bcLK211, and bcLK258, and 4) bcLK232 and bcLK263 (Table S1). Our preliminary paternity analysis, showed a 10% increase of paternity assignment after removing SSR loci with high null allele frequencies, but we included UBCLXtet-21 in spite of its high null allele frequency, because it was easy to multiplex and score as tetra-nucleotide SSR. Additionally, our results showed that the inclusion of this locus did not introduce serious parentage assignment bias. In total, 10 SSR loci were used for parentage assignment (Table S1). After paternity assignment (below), 98% of genotyped offspring were sired by members of the two orchards' panel of fathers. The CERVUS program [31], [32] provides likelihood based paternity inference with a known level of statistical confidence that accounts for genotyping error; we used it to to assign the pollen donor for 1,538 offspring. A genotyping error rate of 0.03 across the 10 loci was estimated from the known mother-offspring genotypes (Table S1). The paternity assignment was based on 10,000 simulations, with the 41-parents as candidate fathers. The log-likelihood (LOD) score, the likelihood that the candidate parent is the true parent divided by the likelihood that the candidate parent is not the true parent, was calculated for each putative parent. The delta score, the difference in LOD scores of the two most likely candidate parents, was used as a criterion for assignment of parentage at the 95% level of confidence in our analysis.

Quantitative genetics analyses

A classical individual-tree additive model, assuming no dominance and epistatic effects, was used. The model included a fixed effect of overall mean (β), a normally distributed random additive genetic effect (a, breeding values), with covariance matrix A = {} where A is the additive relationship matrix (see below [34]) among all trees: parents without records, plus offspring with data, and the additive genetic variance. The model also included a normally distributed random plot effect term (p) with mean zero and variance . Finally, a normally distributed random error (e) with mean zero and variance were included. Let y be a vector containing the tree individual observations for height. Then, in matrix notation, the classical individual-tree additive model can be described as:

$$y = X\beta + Z_p p + Z_a a + e \tag{1}$$

Let A be the additive relationship matrix based on pedigree. The A matrix has diagonal elements equal to 1+Fi, where Fi is the inbreeding coefficient for the ith individual and off diagonals equal to the additive relationships Aij between tree i and j. Three individual-tree additive mixed models (model 91)) were evaluated using different pedigree files. Assuming that parent trees were unrelated, the first model, half-sib (HS model), was used with the known female parent of each individual, where all individuals are assumed not inbred (i.e., Fi = 0), and the additive genetic relationship are 0.25 or 0.0 for both trees with different fathers (with unrelated pollen), thus being maternal half-sibs and unrelated trees, respectively. This model is commonly used by forest geneticists and is called the open-pollinated test, where individuals within an open-pollinated family are assumed to be half-sibs [8]. The pedigree reconstruction created two more scenarios, one includes known female parents for all individuals in the common garden (the sampled 15 seed donors) and the male parents (any one of the orchard's 41 parents) for those used in the pedigree reconstruction (1,419 seedlings) (combined HS+FS model). The second includes only the 1,419 seedlings with their known maternal and paternal parents (known as the FS model). When male parents are known, correct inbreeding coefficient (i.e., Fi = 0.5) and additive relationship between trees ranging from selfs to half-sibs (e.g., Aij = 1 if two individuals are generated by self-pollination or Aij = 0.5 if two individuals are full-sibs through a common father) were considered in the A matrix.

Variance components

Restricted Maximum Likelihood (REML [34]) was used to estimate variances for the random effects of the classical individual-tree additive model (model (1)) and was obtained with the ASReml program [35], which uses the average information algorithm described by Gilmour et al. [36]. The narrow-sense individual heritability (h2) was calculated as , where with i = a, p, and e are the values of the additive, plot, and error variance of the individual-tree model (1). Additionally, the inclusion of male information in the pedigree matrix allowed expanding model (1) to estimate the additive genetic variance after considering the additional genetic relationships generated by pedigree reconstruction. This was done to allow comparing the classical individual-tree additive models used. An important limitation of the REML (co)variance estimates is that their distribution is unknown. Only an approximate measure of precision of the estimates based on asymptotic or large sample theory can be calculated. Approximate standard errors (s.e.) of the and h2 were computed with the "delta method" based on the Taylor expansion [18] using ASREML [35].

Prediction of the breeding values and response to selection

The analysis of a progeny test normally involves two steps: first the estimation of variance components and second the prediction of breeding values for the individuals, using the variance components estimated in the first step. In the three models, the "Best Linear Unbiased Predictors" (BLUPs) of parent and offspring breeding values were computed with ASReml from the estimated variance components. The accuracy of the predicted breeding values was calculated using the following expression: . The acronym PEV stands for 'prediction error variance' [36] of predicted breeding values, using the BLUPs of parent and offspring and Fi is the inbreeding coefficient for the ith individual. The PEV is calculated as the diagonal elements of the inverse of the coefficient matrix from the mixed model equations [36]. To make the accuracies comparable across models (i.e., HS, combined HS+FS and FS), the variance components required to set up the mixed model equations were those estimated from the combined HS+FS. Pearson product-moment correlation and Spearman rank-order correlation were also calculated to compare whether the strength of linear dependence and the ranking of predicated breeding values differed among models. Additionally we have included confidence intervals of all correlation estimates to evaluate jointly the variance and sample size under the alpha value of 0.05. Individual tree BLUP values were used to compare the response to selection under the HS (forward and combined) and combined HS+FS (backward, forward and combined) models, as affected by effective population size, using the optimization protocol outlined in El-Kassaby and Lstibůrek [17].

Estimating offspring optimum sample size

To determinate the optimum number of individuals with known fathers needed for obtaining reliable genetic parameters and thus reducing the DNA fingerprinting efforts, the classical individual-tree additive model (1) was fitted with several pedigree files, where the male information was randomly and progressively deleted, thus increasing percentage of omitted male data from 7 to 92% (i.e., reducing the number of individuals with known male parents). These pedigrees with randomly deleted males provided us with a range of values and standard errors associated with them that the different parameters may take and permitted us to investigate the robustness of results under reduced fingerprinting efforts (i.e., reduce the number of offspring with known paternal parents). For this data set, we set the minimum paternal HS family to n = 6 for inclusion in the analyses and hence the generation of precise genetic parameters and their respective predicted breeding values.

Table S1. Annealing temperature in °C, number of alleles, observed (H_o) and expected (H_e) heterozygosities, and estimated frequencies of null alleles and genotyping error of the seed orchard population (41-Parents).

Locus	Annealing temperature	No. alleles	Heterozygosity H_o	H_e	Null allele Frequency	Genotyping error rate
UBCLXtet_2-11[a]	58	8	0.488	0.528	0.0536	0.0200
UBCLX1-10[a]	58	8	0.700	0.733	0.0183	0.0601
UBCLXtet-21[a]	58	8	0.550	0.810	0.1802	0.1225
UAKLly10a[b]	58	14	0.902	0.877	0.0223	0.0287
UAKLly13-1[b]	58	11	0.683	0.773	0.0557	0.0072
UAKLly13-2[b]	58	7	0.585	0.560	0.0336	0.0092
bcLK066[c]	63 → 53	6	0.846	0.783	0.0459	0.0359
bcLK211[c]	63 → 53	4	0.605	0.598	0.0181	0.0277
bcLK253[c]	63 → 53	9	0.700	0.729	0.0211	0.0033
bcLK258[c]	63 → 53	10	0.703	0.769	0.0456	0.0058
UBCLXA4-1[a]	62	7	0.634	0.740	0.0742	NA[d]
UBCLXdi-16[a]	58	9	0.267	0.812	0.5069	NA
UBCLXdi-21[a]	55	8	0.400	0.698	0.2511	NA
UBCLXtet_2-12[a]	58	4	0.152	0.634	0.6103	NA
bcLK033[c]	63 → 53	6	0.389	0.649	0.2434	NA
bcLK232[c]	63 → 53	8	0.463	0.614	0.1653	NA

bcLK263[c]	63 → 53	9	0.590	0.832	0.1549	NA

[a]Chen *et al.* [28]; [b]Khasa *et al.* [29]; [c]Isoda and Watanabe [30].

[d]NA, data not available for these loci due to their exclusion from paternity assignment.

Annealing temperature in °C, number of alleles, observed (Ho) and expected (He) heterozygosities, and estimated frequencies of null alleles and genotyping error of the seed orchard population used in the present study (41-Parents).

doi:10.1371/journal.pone.0025737.s001

(DOCX)

ACKNOWLEDGMENTS

We are most grateful to D. Reid and C. Walsh for providing seed and reproductive survey data; J. Halusiak for seedling production; UBC graduate students N. Massah, B. Lai, M. Ismail and R. Soolanayakanahally for assistance with the common garden establishment; I. Fundova, T. Funda and B. Lai for trial maintenance and measurements; I. Fundova and C.N. Takuathung for DNA extraction, L. Bouffier, R.J. Peti, M. Stoehr and P.E. Smouse for critical and constructive review of earlier draft.

Author Contributions

Conceived and designed the experiments: YAE. Performed the experiments: YAE CL. Analyzed the data: YAE EPC CL JK ML. Wrote the paper: YAE.

REFERENCES

1. Allard RW (1960) Principles of plant breeding (John Wiley and Sons, YN).
2. Namkoong G, Kang HC, Brouard JS (1988) Tree breeding: principles and strategies (Springer-Verlag, NY, Monograph, Theor Appl Genet 11).

3. Falconer DS, Mackay TFC (1996) Introduction to quantitative genetics (Longman, NY).
4. White TL, Adams WT, Neale DB (2007) Forest genetics (CABI Publishing, Cambridge, MA).
5. Stern K, Roche L (1974) Genetics of forest ecosystems (Chapman and Hall, London).
6. Cotterill PP (1986) Genetic gains expected from alternative breeding strategies including simple low cost options. Silvae Genet 35: 212–223.
7. Burdon RD, Shelbourne CJA (1971) Breeding populations for recurrent selection: conflicts and possible solutions. NZ J For Sci 1: 174–193.
8. Namkoong G (1966) Inbreeding effects on estimation of genetic additive variance. For Sci 12: 8–13. View Article PubMed/NCBI Google Scholar
9. Squillace AE (1974) Average genetic correlations among offspring from open-pollinated forest trees. Silvae Genet 23: 149–156.
10. Askew GR, El-Kassaby YA (1994) Estimation of relationship coefficients among progeny derived from wind-pollinated orchard seeds. Theor Appl Genet 88: 267–272.
11. Lambeth C, Lee BC, O'Malley D, Wheeler N (2001) Polymix breeding with parental analysis of progeny: an alternative to full-sib breeding and testing. Theor Appl Genet 103: 930–943.
12. Grattapaglia D, Ribeiro VJ, Rezende GDSP (2004) Retrospective selection of elite parent trees using paternity testing with microsatellite markers: an alternative short term breeding tactic for Eucalyptus. Theor Appl Genet 109: 192–199.
13. El-Kassaby YA, Lstibůrek M, Liewlaksaneeyanawin C, Slavov GT, Howe GT (2006) Breeding without breeding: approach, example, and proof of concept. pp. 43–54.
14. Gaspar MJ, de-Lucas AI, Alia R, Paiva JAP, Hidalgo E, et al. (2009) Use of molecular markers for estimating breeding parameters: a case study in a Pinus pinster Ait. progeny trial. Tree Genet Genomes 5: 609–616.
15. Hansen OK, McKinney LV (2010) Establishment of a quasi-field trial in Abies nordmanniana – test of a new approach to forest tree breeding. Tree Genet Genomes 6: 345–355.
16. Doreksen TK, Herbinger CM (2010) Impact of reconstructed pedigrees on progeny-test breeding values in red spruce. Tree Genet Genomes. DOI 10.1007/s11295-010-0274-1.
17. El-Kassaby YA, Lstibůrek M (2009) Breeding without breeding. Genet Res 91: 111–120.

18. Lynch M, Walsh B (1998) Genetics and analysis of quantitative traits (Sinauer Associates, Sunderland, MA).
19. Funda T, Liewlaksaneeyanawin C, Fundova I, Lai BSK, Walsh C, et al. (2011) Congruence between clonal reproductive investment and success as revealed by DNA-based pedigree reconstruction in seed orchards of lodgepole pine, Douglas-fir, and western larch. Can J For Res 41: 380–389.
20. Ritland K (1996) A marker-based method for inferences about quantitative inheritance in natural populations. Evolution 50: 1062–1073.
21. Ritland K, Ritland C (1996) Inferences about quantitative inheritance based upon natural population structure in the common monkeyflower, Mimulus guttatus. Evolution 50: 1074–1082.
22. Ritland K, Travis S (2004) Inferences involving individual coefficients of relatedness and inbreeding in natural populations of Abies. Fort Ecol Manage 197: 171–180.
23. Namkoong G (1979) Introduction to quantitative genetics in forestry (US Depart Agriculture, Forest Service, Washington, DC, Tech Bulletin No 1588).
24. Rönningen K, Van Vleck LD (1985) pp. 187–225. General and quantitative genetics (World Animal Science, Elsevier, NY).
25. Cappa EP, Lstiburek M, Yanchuk AD, El-Kassaby YA (2011) Two-dimensional penalized splines via Gibbs sampling to account for spatial variability in forest genetic trials with small amount of information available. Silvae Genet 60: 25–35.
26. El-Kassaby YA, Barnes S, Cook C, MacLeod DA (1993) Supplemental-mass-pollination success rate in a mature Douglas-fir seed orchard. Can J For Res 23: 1096–1099.
27. International Seed Testing Association (1993) International rules for seed testing. Seed Sci Technol 21S 284.
28. Chen CC, Liewlaksaneeyanawin C, Funda T, Kenawy AMA, Newton CH, et al. (2008) Development and characterization of microsatellite loci in western larch (Larix occidentalis Nutt.). Mol Ecol Res 9: 843–845.
29. Khasa PD, Newton C, Rahman M, Jaquish B, Dancik BO (2000) Isolation, characterization and inheritance of microsatellite loci in alpine larch and western larch. Genome 43: 439–448. View Article PubMed/NCBI Google Scholar
30. Isoda K, Watanabe A (2006) Isolation and characterization of microsatellite loci from Larix kaempferi. Mol Ecol Notes 6: 664–666.

31. Marshall TC, Slate J, Kruuk LEB, Pemberton JM (1998) Statistical confidence for likelihood-based paternity inference in natural populations. Mol Ecol 7: 639–655.

32. Kalinowski ST, Taper ML, Marshall TC (2007) Revising how the computer program CERVUS accommodates genotyping error increases success in paternity assignment. Mol Ecol 16: 1099–1106.

33. Dakin EE, Avise JC (2004) Microsatellite null alleles in parentage analysis. Heredity 93: 504–509.

34. Henderson CR (1984) Applications of linear models in animal breeding (University of Guelph, Guelph, ON, Canada).

35. Gilmour AR, Gogel BJ, Cullis BR, Thompson R (2006) ASReml user guide (Release 2.0 VSN International, Hemel Hempstead, UK).

36. Gilmour AR, Thompson R, Cullis BR (1995) Average information REML, an efficient algorithm for variance parameter estimation in linear mixed models. Biometrics 51: 1440–1450.

Chapter 5

TRANSCRIPTOME ANALYSIS OF DIFFERENTIALLY EXPRESSED GENES RELEVANT TO VARIEGATION IN PEACH FLOWERS

[1]Yingnan Chen, [1]Yan Mao, [1]Hailin Liu, [2]Faxin Yu, [1]Shuxian Li, [1]Tongming Yin

[1]Yingnan Chen, [1]Yan Mao, [1]Hailin Liu, [2]Faxin Yu, [1]Shuxian Li, [1]Tongming Yin

[1]The Southern Modern Forestry Collaborative Innovation Center, Nanjing Forestry University, Nanjing, Jiangsu, China

[2]Institute of Biology and Resources, Jiangxi Academy of Sciences, Nanchang, Jiangxi, China

ABSTRACT

Background

Variegation in flower color is commonly observed in many plant species and also occurs on ornamental peaches (Prunus persica f. versicolor [Sieb.] Voss). Variegated plants are highly valuable in the floricultural market. To gain a global perspective on genes differentially expressed in variegated peach flowers, we performed large-scale transcriptome sequencing of white and red petals separately collected from a variegated peach tree.

Results

A total of 1,556,597 high-quality reads were obtained, with an average read length of 445 bp. The ESTs were assembled into 16,530 contigs and 42,050 singletons. The resulting unigenes covered about 60% of total predicted genes in the peach genome. These unigenes were further subjected to functional annotation and biochemical pathway analysis. Digital expression analysis identified a total of 514 genes differentially expressed between red and white flower petals. Since peach flower coloration is determined by the expression and regulation of structural genes relevant to flavonoid biosynthesis, a detailed examination detected four key structural genes, including C4H, CHS, CHI and F3H, expressed at a significantly higher level in red than in white petal. Except for the structural genes, we also detected 11 differentially expressed regulatory genes relating to flavonoid biosynthesis. Using the differentially expressed structural genes as the test objects, we validated the digital expression results by using quantitative real-time PCR, and the differential expression of C4H, CHS and F3H were confirmed.

Conclusion

In this study, we generated a large EST collection from flower petals of a variegated peach. By digital expression analysis, we identified an informative list of candidate genes associated with variegation in peach flowers, which offered a unique opportunity to uncover the genetic mechanisms underlying flower color variegation.

Background

Flower color is of paramount importance in plant biology [1]. Three major groups of pigments–betalains, carotenoids and flavonoids–are responsible for the attractive natural display of flower colors [2], [3]. Flavonoids, particularly anthocyanidins, are the most common flower pigments, and contribute to a wide range of colors, from pale yellow to red, purple and blue [4]. To date, most enzymes involved in the anthocyanin biosynthetic pathway have been identified in various plant species [5]–[7]. All anthocyanidins are derived from a general phenylpropanoid pathway that converts the aromatic amino acid phenylalanine to anthocyanidins through a series of enzyme-catalyzed reactions [8]–[12]. During the past few decades, much of the molecular information available on the regulation and biosynthesis of flower pigments has been derived from studies performed in model systems such as Zea mays L. (maize) [13], [14], Arabidopsis [6], [15], petunia [16], [17] and snapdragon [3], [15]. An increasing number of non-classical plants, however, are providing unique insights into molecular mechanisms involved in flower pigment formation, leading to further understanding of how flower color varies among wild species [18]–[20].

Ornamental peach, a member of the Rosaceae family, is an important horticultural tree. Prunus persica f. versicolor (Sieb.) Voss, one of the main varieties, is characterized by the presence of chimeric flowers (Figure 1). Because variegation in flowers often attracts consumer attention, variegated plants are generally of high value in the ornamental market. This unstable phenotype has been observed in natural populations of petunia, snapdragon, morning glory, azalea and other plant species [21]. Flower variegation is usually due to the presence of a group of colored cells descended from a single ancestral cell in which a somatic mutation from the recessive white to the pigmented revertant allele has occurred. Somatic mutation frequency and timing during petal development determine variegation patterns [22].

Figure 1. Sampling stage of variegated flower petals.

doi:10.1371/journal.pone.0090842.g001

Molecular mechanisms of flower variegation have been investigated in various plant species [22]–[25]. Some of this variegation is caused by transposable element insertion into structural genes associated with the anthocyanin synthetic pathway [26]–[28]. Because of the complexity of the flavonoid biosynthetic pathway, however, the exact genetic mechanisms underlying the generation of flower pigmentation chimeras may differ among species. Chaparro et al. [29] studied variegation in anthocyanin production in both vegetative and reproductive tissues of the peach cultivar Pillar; they verified that this phenotype is heritable, although the degree of variegation differed according to the genetic background of outcross progeny. The genetic mechanisms in peach responsible for this unstable phenotype have not yet been elucidated. The whole peach genome has been sequenced [30], and abundant transcriptome sequences are available to the public [31]. The reference genome and large EST collection enhance our ability to align sequences, identify genes, and characterize transcriptomes, thereby facilitating identification of the genetic basis of variegated pigmentation in peach. To understand the mechanism of variegated pigmentation, detection of differentially expressed genes from different-colored flowers is essential. Transcriptome sequencing is an efficient way to measure transcriptome composition and uncover differentially expressed genes [32]–[34]. Many studies using high-throughput next-generation sequencing technology have surveyed the complex transcriptomes of various plants including Arabidopsis thaliana [35], Digitalis purpurea [36], Carthamus tinctorius [37], Persea americana [38] and Salix suchowensis [39]. In this study, transcriptomes of white and red flower petals sampled from a Prunus persica f. versicolor individual were sequenced using a 454 GS-FLX sequencer. By analyzing the data with various bioinformatics tools, we aimed to discover candidate genes involved in peach flower variegation.

Materials and Methods

Plant Material

We separately collected petals of expanded but unflushed white and red variegated flower buds (Figure 1) from a tree of Prunus persica f. versicolor in Nanjing Lovers Garden, Jiangsu, China in March 2012. Petals were immediately frozen in liquid nitrogen and stored at −80°C until RNA extraction. The field studies did not involve any endangered or protected species, and sample collection was authorized by the administration office of Nanjing Lovers Garden.

RNA Extraction and cDNA Synthesis

Total RNA was extracted separately from white and red petals using the CTAB method [39]. RNA integrity was confirmed by 1% agarose gel electrophoresis. After digestion with DNase (Takara) at 37°C for 30 min to remove DNA residues, RNA concentration was determined using a Nanodrop spectrophotometer (Thermo). mRNA was then purified from total RNA using an Oligotex mRNA midi kit (Qiagen), with its quality assessed using an Agilent Technologies 2100 Bioanalyzer (Agilent). cDNA synthesis was performed using the 454 cDNA amplification technique with a cDNA Synthesis System kit (Roche) following the manufacturer's protocol.

Sequencing Library Construction and 454 Sequencing

Sequencing libraries were separately constructed for white and red petals using a Rapid Library Prep kit (Roche). Quality of sequencing libraries was checked using an Agilent 2100 Bioanalyzer. Approximately 2.1 million beads per flower color were separately loaded onto two sections of a pico-titer plate. A sequencing run was carried out on a Roche 454 GS FLX sequencer at Nanjing Forestry University. All ESTs in this study were deposited in NCBI with an accession number SRR1037160.

EST Sequence Processing and Assembly

Raw 454 sequence files in SFF format were base-called using the Pyrobayes base caller [40]. In addition, 78,689 peach ESTs were downloaded from GenBank in February 2012. The 454 sequencing data and GenBank ESTs were processed with GS FLX v2.0.01 software (454 Life Sciences, Roche) to remove low-quality and adaptor sequences. To remove possible contamination, the resulting high-quality 454 and GenBank sequences were screened against the NCBI UniVec database, E. coli genome sequences, and peach ribosomal RNA and chloroplast genome sequences. Sequences shorter than 50 bp were discarded before assembly. Finally, the processed 454 and GenBank sequences were assembled into putative transcripts (including contigs and singletons) using the 454 assembly program Newbler v2.7 with the following overlap detection settings: seed step, 12; seed length, 16; seed count, 1; minimum overlap length, 40; minimum overlap identity, 95%; alignment identity score, 2; alignment difference score, −3. We parsed the 454 ReadStatus.txt file to identify singletons, which were unassembled reads. The contig and singleton files were used to generate a unigene file.

Mapping Unigenes to Peach Genome Predicted Genes

The generated unigenes were aligned to predicted genes of the peach genome using BLAST [41]. Based on the International Peach

Genome Initiative (IPGI; http://www.rosaceae.org/species/prunus/prunus_persica), there are 27,864 predicted genes in the peach genome. Because the number of obtained unigenes was much greater than the number of predicted genes, it is possible that different unigenes were segments of the same predicted gene. In this study, unigenes mapping to the same predicted gene were integrated into one unique gene, which was used to collect transcript count information for detection of differentially expressed genes in white and red flower petals.

Gene Annotation and Pathway Prediction

Unigene annotation was performed by BLASTX analysis [42] against peach protein (http://www.phytozome.com/peach.php; v2.0), NCBI non-redundant protein (nr) and UniProt databases using a cutoff E-value of 10−5. Gene Ontology (GO) terms were assigned to each unigene based on GO annotations of its corresponding homologs in the UniProt database [43], and, using interpro2go and pfam2go mapping files available on the GO website (http://www.geneontology.org), their corresponding InterPro and Pfam domains. GO mapping results were further plotted by uploading the GO list file to the Web Gene Ontology Annotation Plot (WEGO) website (http://wego.genomics.org.cn/cgi-bin/wego/index.pl). The detailed annotation was then used to retrieve keywords for identification of flower pigmentation-related genes.

The KEGG Automatic Annotation Server (KAAS; http://www.genome.jp/tools/kaas/) was employed to perform metabolic pathway mapping [44], [45]. KAAS assigned each peach gene a KEGG Orthology (KO) number, which was then used for mapping to a KEGG reference metabolic pathway.

Differentially Expressed Genes between White and Red Petals

Following cDNA sequence assembly and gene prediction, transcript count information was collected for sequences corresponding to each predicted gene associated with a flower petal color. To obtain relative expression levels in each sample, transcript counts were normalized to the total number of produced transcripts per sample. Significance of gene differential expression level was assessed using R [46], $\chi 2$ and Fisher exact tests as implemented in the publicly available web tool IDEG6 (http://telethon.bio.unipd.it/bioinfo/IDEG6/) [47]. A gene was considered to be differentially expressed when results from the above tests were all significant at a level of $P \leq 0.0001$.

Quantitative Real-time PCR Analysis

Total RNA was extracted from red and white flower petals as described above. Approximately 2 μg of total RNA per sample was treated with DNaseI (Takara), and then subjected to reverse transcript to cDNA using reverse transcription system (Promega). The quantitative real-time PCR (qRT-PCR) was performed using Applied Biosystems 7500 Real-time PCR system (Applied Biosystems) with SYBR Premix Ex Taq (Takara). Each reaction contained 2 μL the first-strand cDNA as template, in a total volume of 20 μL reaction mixture. The amplification program was performed as 95°C 30 s followed by 95°C for 5 s and 60°C for 34 s (40 cycles). Gene-specific primers, shown in Table S3, were used for detecting the relative quantification of each gene. qRT-PCR expression levels were compared based on the mean of three independent experimental repeats. Calculation of relative expression level was performed using the $2-\Delta\Delta CT$ method [48]. TEF2 was used as an internal control for normalization [49].

RESULTS

Transcriptome Sequencing and Assembly

A half 454 GS-FLX run was performed for each sample, resulting in the generation of 1,556,684 reads. After quality control, 1,556,597 reads with an average length of 1, 445 bp remained: 837,041 from white petals and 719,556 from red petals (Table 1). These reads, together with 76,822 high-quality GenBank ESTs, were subjected to assembly analysis. Sequence assembly yielded 58,580 unigenes comprising 16,530 contigs and 42,050 singletons. Contigs and singletons had average lengths of 1,555 bp and 364 bp, respectively (Table 2). A plot of unigene length distribution revealed that most unigenes (82.8%)

were longer than 300 bp (Figure 2). From our EST collection, we were able to identify a number of highly expressed unigenes in peach flowers (Figure 3). Approximately 12,589 unigenes were assembled from more than 10 EST reads; these unigenes (~21% of all unigenes) corresponded to ~95% of total EST reads.

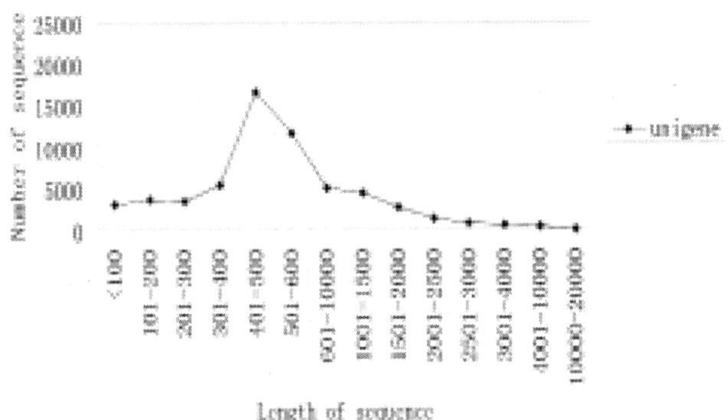

Figure 2. Length distributions of peach unigenes.

doi:10.1371/journal.pone.0090842.g002

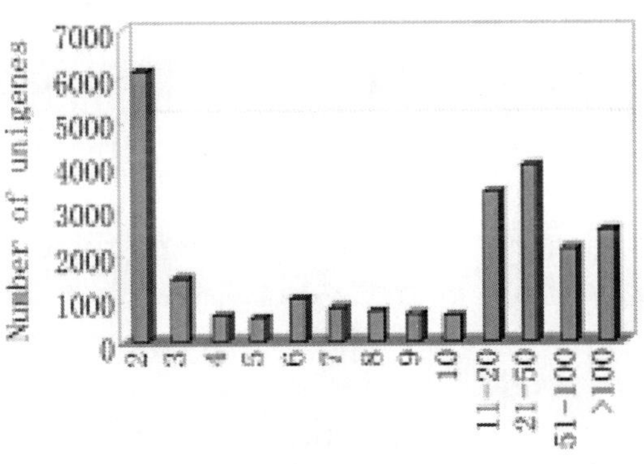

Figure 3. Distribution of ESTs in peach unigenes.

doi:10.1371/journal.pone.0090842.g003

Table 1. Statistics for peach ESTs generated by 454 GS-FLX sequencing.

doi:10.1371/journal.pone.0090842.t001

	Red petals	White petals	Total
Number of reads	719,556	837,041	1,556,597
Average read length (bp)	443.3	448	445.8
Total bases (bp)	318,983,491	374,962,455	693,945,946
Number of reads in contigs	675,990	783,159	1,459,149
Number of reads as singletons	28,444	35,141	63,585

doi:10.1371/journal.pone.0090842.t001

Table 2. Statistics for assembled peach unigenes.

doi:10.1371/journal.pone.0090842.t002

	Singleton	Contig	Unigene(cluster)
Number of sequences	42,050	16,530	58,580
Average read length (bp)	364	1,555	699
Total bases (bp)	26,469,716	27,763,950	40,933,465
Number of unigenes mapped to peach genome predicted genes	25,042	15,645	40,687
Number of unigenes unmapped to peach genome predicted genes	17008	885	17893

Mapping Unigenes to Predicted Genes in the Peach Genome

Using the peach genome (v1.0; IPGI), approximately 70% (40,687) of obtained unigenes could be mapped to predicted genes of P. persica. Among these, 15,645 were contigs (95% of all contigs) and 25,042 were singletons (60% of singletons). The mapped unigenes were aligned to 16,733 predicted genes, corresponding to approximately 60% of the 27,864 predicted genes in the peach genome. The remaining 17,893 unigenes could not be mapped to any predicted genes. Among the unmappable unigenes, 885 were contigs and 17,008 were singletons (Table 2). We further examined the unmappable unigenes, and found that about 20.6% had no significant matches with plant sequences. In total, alignment of unigenes to predicted P. persica genes generated 34,626 unique genes: 16,733 identified as peach genome predicted genes and 17,893 unmappable unigenes. When unigenes were mapped to different scaffolds of the peach genome (http://www.rosaceae.org/species/prunus/prunus_persica), the number of mapped unigenes was found to be highly correlated with the number of predicted genes on each scaffold (paired t-test, R2 = 0.992) (Figure 4). The number of expressed genes was thus proportional to the number of predicted genes allocated to each scaffold.

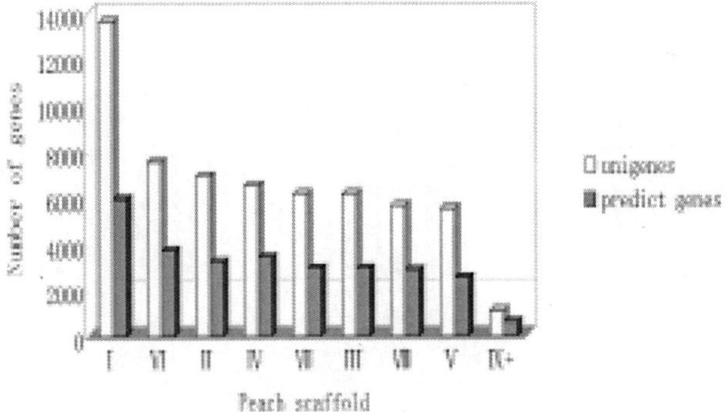

Figure 4. Distribution of unigenes and predicted genes allocated to each peach genome scaffold.

Roman numerals I–VIII correspond to the eight large scaffolds of the peach genome (v1.0; IPGI), and are ordered along the x-axis with the number of mapped unigenes on each scaffold. IX+ refers to all remaining small scaffolds of the peach genome.

doi:10.1371/journal.pone.0090842.g004

Functional Annotation of Peach Transcriptomes

Based on alignments of unigenes to peach genome predicted genes, 34,626 unique genes were identified. To infer putative functions of these unique genes, we blasted their sequences against the GenBank nr database using a significance cutoff of E ≤10−5. The analysis indicated that 21,663 unique genes (63%) had significant matches to the nr database, among which 15,846 were peach genome predicted genes (94.7% of the 16,733 unigene-matched predicted genes) and 5,817 were unmappable unigenes (32% of all unmappable unigenes). Most of the unmappable unigenes were singletons (95.1%), which were much shorter than contigs (364 bp vs. 1,555 bp on average). In general, the longer the sequence is, the greater the chance of annotation, and thus the more number of GO terms that can be recovered [50]. This suggests that the inability to assign putative functions to most unmappable unigenes was due to the absence of conserved functional domains in short sequences.

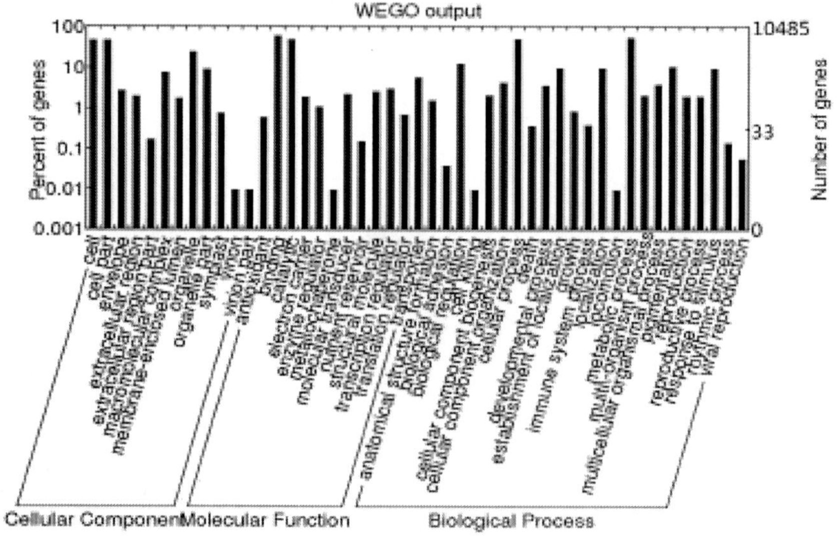

Figure 5. Number of peach unigenes in each functional category.

Peach unigenes were classified into different functional groups based on a set of plant-specific GO slims within cellular component, molecular function and biological process categories.

doi:10.1371/journal.pone.0090842.g005

Gene Ontology (GO) describes gene products in terms of their associated molecular functions, biological processes and cellular components. To assign putative functional roles to the obtained unique genes, GO terms were assigned based on sequence similarities to known GO-annotated proteins (and their InterPro and Pfam domains) in the UniProt database. As a result, 10,485 unique genes were assigned at least one GO term. Among GO terms, 896 were in the biological process category, 606 in the molecular function category, and 607 in the cellular component category (Figure 5). These unique genes were further displayed using a set of GO slims, which are a list of high-level GO terms providing a broad overview of ontology content (http://www.geneontology.org/GO.slims.shtml). Figure 5 displays the functional classification of peach unique genes into plant-specific GO slims within cellular component, molecular function and biological process categories. Genes involved in cell, cell part, binding, catalytic, cellular process and metabolic process categories were the highest-represented groups, indicating that flower buds were undergoing rapid growth and carrying out intensive metabolic activities. In the biological process category, it is noteworthy that genes involved in pigmentation were also highly represented, indicating active pigmentation activities.

Biochemical Pathways

KEGG, an alternative functional gene annotation system, performs assignments based on Enzyme Commission (EC) numbers of genes associated with biochemical pathways. To further demonstrate the usefulness of the generated peach ESTs for discovering flower pigmentation-related genes, we identified biochemical pathways represented by our EST collection. Annotations of peach unique genes were fed into the Pathway Tools program (KAAS; http://www.genome.jp/tools/kaas/). This process predicted 229 pathways represented by a total of 5,345 unique genes. Of these KEGG-annotated genes, 2,425 (45.4%) were involved in metabolism, 1,186 (22.2%) in genetic information processing, 346 (6.5%) in environmental information processing, 569 (10.6%) in cellular processes and 792 (14.8%) in organism systems (Table S1). Among genes involved in metabolism, 14 genes were found to encode key enzymes involved in flavonoid biosynthesis. We mapped and highlighted these genes onto the "flavonoid biosynthesis" pathway (Figure S1), which demonstrated that most of the key enzymes in this pathway were covered by our sequence data. We also detected three genes involved in flavone and flavonol biosynthesis; they were mapped and highlighted onto the "flavone and flavonol biosynthesis" pathway (Figure S2). In contrast to the flavonoid biosynthesis pathway, only a small proportion of flavone and flavonol biosynthetic pathway key enzymes were represented by our sequence data.

Identification of Differentially Expressed Genes in White and Red Flower Petals

In this study, we obtained 16,530 contigs and 42,050 singletons, among which 15,645 contigs and 25,042 singletons were mapped to 16,733 peach predicted genes. There were 855 unmapped contigs and 17,008 unmapped singletons. The purpose of this study was to uncover genes differentially expressed between flower petal colors. The 17,008 unmapped singletons were less closely related to peach genome predicted genes and were not present in sufficient quantities for statistical analysis [50]; they were consequently excluded from the differentially expressed gene analysis. Number of reads was obtained for each gene using a custom PERL script. The digital expression profiling analysis identified 514 genes differentially expressed between red and white flower petals, with $P<0.0001$ for all employed statistics (Table S2); 367 of these genes showed significantly higher expression in red flower petals, whereas 147 exhibited significantly higher expression in white petals.

In peach, the flavonoid biosynthetic pathway leads to production of colored pigments. We therefore specifically examined expression of flavonoid structural genes, and identified four differentially expressed genes, all highly expressed in red petals. These genes encode the enzymes chalcone and stilbene synthase (CHS), chalcone-flavanone isomerase (CHI), cinnamate-4-hydroxylase (C4H) and flavanone 3-hydroxylase (F3H). Functionally, C4H catalyzes the incorporation of a 4′-hydroxyl group into cinnamate during p-coumarate formation. CHS condenses one molecule of p-coumaroyl-CoA with three molecules of malonyl-CoA to produce the chalcone tetrahydroxychalcone. (Chalcone is the precursor for all classes of flavonoids, including flavones, flavonols, flavandiols, flavan-4-ols, condensed tannins, isoflavonoids and anthocyanins.) CHI is responsible for the conversion of tetrahydroxychalcone to naringenin, and F3H catalyzes the formation of DHK from naringenin (Figure S3). DHK can be further hydroxylated to form dihydroquercetin (DHQ) and dihydromyricetin (DHM). DHK, DHQ and DHM subsequently lead synthetic branches producing pelargonidin-based (orange to red), cyanidin-based (red to magenta) and delphinidin-based (purple) pigments, respectively. Interestingly, the four differentially expressed genes all mapped to early committed steps on the flavonoid biosynthetic pathway before the formation of DHK (Figure S3). From 454 sequencing, absolute read numbers of C4H, CHS, CHI and F3H in red vs. white flower petals were 146:54, 4050:1448, 122:14 and 26:1, respectively. After normalization, expression levels of C4H, CHS, CHI and F3H were 3.1-, 3.3-, 10.1- and 30.2-fold higher, respectively, in red flower petals than in white ones.

In the flavonoid biosynthetic pathway, transcription levels of flavonoid biosynthesis genes are regulated by various TFs. TFs related to flavonoid biosynthesis can be divided into three classes: MYB, basic-Helix-Loop-Helix (bHLH) and WD40 [15], [51], [52]. Apart from the structural genes, we further examined expression levels of these three TF classes. We detected 11 TFs differentially expressed between red and white petals, including 3 bHLHs, 4 MYBs and 4 WD40s. Nine of these TFs were highly expressed in red petals, and two were highly expressed in white petals (Table 3). These DNA-binding proteins interact with promoter regions of target genes and regulate the initiation rate of mRNA synthesis.

Table 3. Differentially expressed structural and regulatory genes related to flavonoid pathways in variegated peach flowers.

doi:10.1371/journal.pone.0090842.t003

ID	Function distribution	Highly expressed tissue
ppa025745m	Chalcone and stilbene synthase (CHS) family protein	red
ppa011276m	Chalcone-flavanone isomerase (CHI) family protein	red
ppa004544m	cinnamate-4-hydroxylase (C4H)	red
ppa007636m	flavanone 3-hydroxylase (F3H)	red
ppa009757m	basic helix-loop-helix (bHLH) DNA-binding superfamily protein	red
ppa003543m	basic helix-loop-helix (bHLH) DNA-binding superfamily protein	red
ppa013751m	LI1 binding bHLH 1	red
ppa010069m	myb domain protein 113	red
ppa011751m	myb domain protein 24	red
ppa010277m	myb domain protein 4	red
ppa007222m	myb-like transcription factor family protein	white
ppa005673m	Transducin family protein/WD-4 repeat family protein	red
ppa005800m	Transducin family protein/WD-4 repeat family protein	red
ppa002072m	Transducin/WD4 repeat-like superfamily protein	red
ppa014936m	Transducin/WD4 repeat-like superfamily protein	white

doi:10.1371/journal.pone.0090842.t003

Using the differentially expressed structural genes as the test objects, we further validated the digital expression profiling by qRT-PCR technology. Statistically, C4H, CHS, and F3H expressed significantly higher in red flower petals than in white ones, with p<0.05 (Figure 6). Thus, the significant difference in expression level of these three structural genes was confirmed by both of the analytical techniques. Whereas the expression level of CHI was not statistically different between colors. The comparative transcription level of this gene was found to be only slightly higher in red than in white petals by qRT-PCR analysis (Figure 6). Nevertheless, in general, the qRT-PCR results were in rough accordance with the electronic data of gene expression analysis.

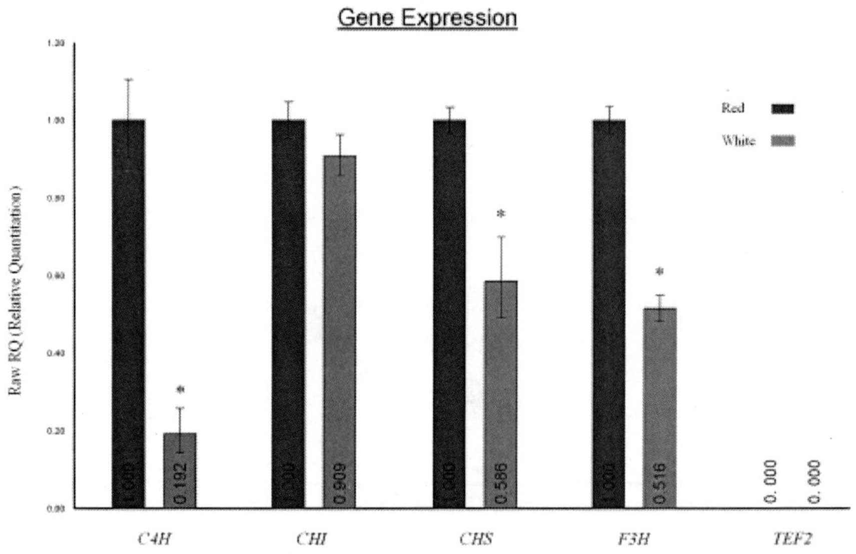

Figure 6. qRT-PCR analysis of C4H, CHS, CHI and F3H gene expression in white and red flower petals.

TEF2 was used as an internal control for normalization. qRT-PCR data calculated with the 2−ΔΔCt method [48]. The expression level of each gene in red flower petals was arbitrarily set as 1 and its corresponding transcript level in white flower petals was calibrated against red one. Error bars represent standard error for three independent experimental replicates. *, p<0.05, as determined by one-way ANOVA.

doi:10.1371/journal.pone.0090842.g006

Discussion

Approximately 60% of total predicted genes in the peach genome were covered by the unigenes obtained in this study. In Arabidopsis, approximately 55–67% of genes are expressed in a single tissue based on microarray analysis [53]. In human and mouse, around 60–70% of genes are expressed in a specific tissue [54]. Transcriptomic studies have revealed that about 64%, 66% and 68% of genesare expressed in whole flower tissues of cucumber [55], willow [39] and peach [31], respectively. In our study, we sequenced genes expressed in flower petals, not entire flowers. Our EST collection, designed to capture the majority of genes expressed in peach flower petals, would thus be expected to encompass fewer expressed genes. Because more transcript sequences were generated in peach (1.5 million) than in willow (1.2 million) [39] or cucumber (0.35 million) [55], the sequence depth of genes covered by our EST dataset should therefore be higher. We thus obtained a reliable dataset to explore differentially expressed genes relevant to variegation in peach flowers. Most differentially expressed genes that have currently been revealed by transcriptome sequencing are derived from studies performed with tissues of different genotypes [39], [55], [56]. In contrast, ESTs generated in our study are collections of expressed genes of the same genotype. Our approach is thus appropriate for identification of differentially expressed genes relevant to the focal phenotype.

The primary pigments related to flower color are anthocyanins, which contribute to a variety of colors, such as red, pink and blue [4]. Biochemical pathway mapping revealed that most of the structural genes in the anthocyanin biosynthetic pathway were covered by our sequence data (Figure S1), and we detected four structural genes highly expressed in red petals. In addition to anthocyanins, some co-pigments, such as flavones and flavonols, can change flower color hues. Only a few structural genes (Figure S2) involved in flavone and flavonol biosynthesis were captured, however, and none of them were differentially expressed between red and white petals. These results suggest that peach flower variegation is linked mainly to anthocyanin biosynthesis, with flavones and flavonols having a limited role.

With respect to enzymatic genes involved in the anthocyanin biosynthetic pathway, C4H, CHS and F3H were confirmed to express at a significantly higher level in red petals than in white petals by both the transcriptome profiling and qRT-PCR analysis. These genes were associated with committed steps before the formation of DHK (Figure S3). Because red flower color in peach is related to the synthesis of pelargonidin-based (orange to red) and cyanidin-based (red to magenta) pigments, our results suggest that the low C4H, CHS and F3H expression levels in white petals reduce DHK formation, thereby inhibiting pelargonidin and cyanidin production. In contrast, the high expression levels of these genes observed in red petals ensure sufficient anthocyanin yields to make flowers red. In this experiment, none of the structural genes downstream of DHK (Figure S3) were found to be significantly differentially expressed between white and red petals. It can thus be concluded that variegation in peach flowers is related primarily to structural genes upstream of DHK in the anthocyanin synthetic pathway, and is less likely affected by genes downstream.

In nature, white is the most abundantly occurring flower color, and can also be artificially produced from colored flowers. The first successful reversion to white flowers was achieved by suppressing CHS in tobacco and petunia [57]. Suppression of structural genes in the anthocyanin synthetic pathway has subsequently proven useful for modifying colored flowers to white. Studies have shown that transgenic plants carrying single gene constructs of CHS [58], [59] or F3H [60] can all exhibit white or faint-colored flowers.

Final anthocyanin concentrations in plant cells are not determined solely by structural gene expression levels; some regulatory genes are clearly also involved in control of flavonoid biosynthesis gene expression. These regulatory genes, typically specific TFs, influence anthocyanin biosynthesis intensity and pattern and generally control expression of many different structural genes [21]. Three classes of TFs–bHLH, MYB and WD40–have been found to be related to flavonoid biosynthesis [15], [51], [52]. In maize, an MYB-related protein and a bHLH-containing protein have been shown to interact to activate genes in the anthocyanin biosynthetic pathway [61]. In this study, we detected 11 differentially expressed TFs from the three flavonoid biosynthesis-related classes. The high C4H, CHS and F3H expression levels observed in red peach flower petals may be due to regulation by one or more of these TFs. Although the exact regulatory TFs remain unknown, this study provides an informative list of candidates. In peach flowers, different variegation patterns are observed: some occur among different flowers on the same branch, while others occur within the same flower (Figure S4). Much genetic evidence supports the hypothesis that flower variegation is caused by transposable elements inserted into structural genes or regulatory elements related to anthocyanin synthesis [26]–[28]. The different transcript levels detected in this study for structural genes and TFs might be due to insertions of transposable elements. Examination of candidate DNA sequences for transposable element insertions is needed to determine the mechanisms of peach flower variegation. Although transposable elements cannot be detected based solely on transcriptome sequencing, our study has provided some novel insights into the molecular mechanisms underlying variegation in peach flowers.

CONCLUSIONS

Flower color is one of the most attractive characteristics of ornamental plants, and variegated plants are highly valuable in the floricultural market. In this study, we generated a large EST collection from flower petals of a variegated peach. Based on the digital expression analysis, a total of 514 genes were identified to differentially express between red and white flower petals. The red and white petals were collected from the same tree, and they were in the same developmental stage. Moreover, flower coloration was specifically connected with the flavonoid biosynthetic pathway. All these conditions enabled us to narrow down the considerable number of differentially expressed genes to a small number of candidates, which warranted further investigation. Finally, three key structural genes in the anthocyanin biosynthesis pathway were confirmed to express significantly different between colors, and we also detected 11 differentially expressed TFs related to anthocyanin biosynthesis. Our results provide critical information for uncovering candidate genes associated with variegation in peach flowers. We believe this transcriptome dataset will continue to provide unique insights into the molecular mechanisms controlling variegated flower pigmentation, and will eventually help the molecular engineering of variegated plants.

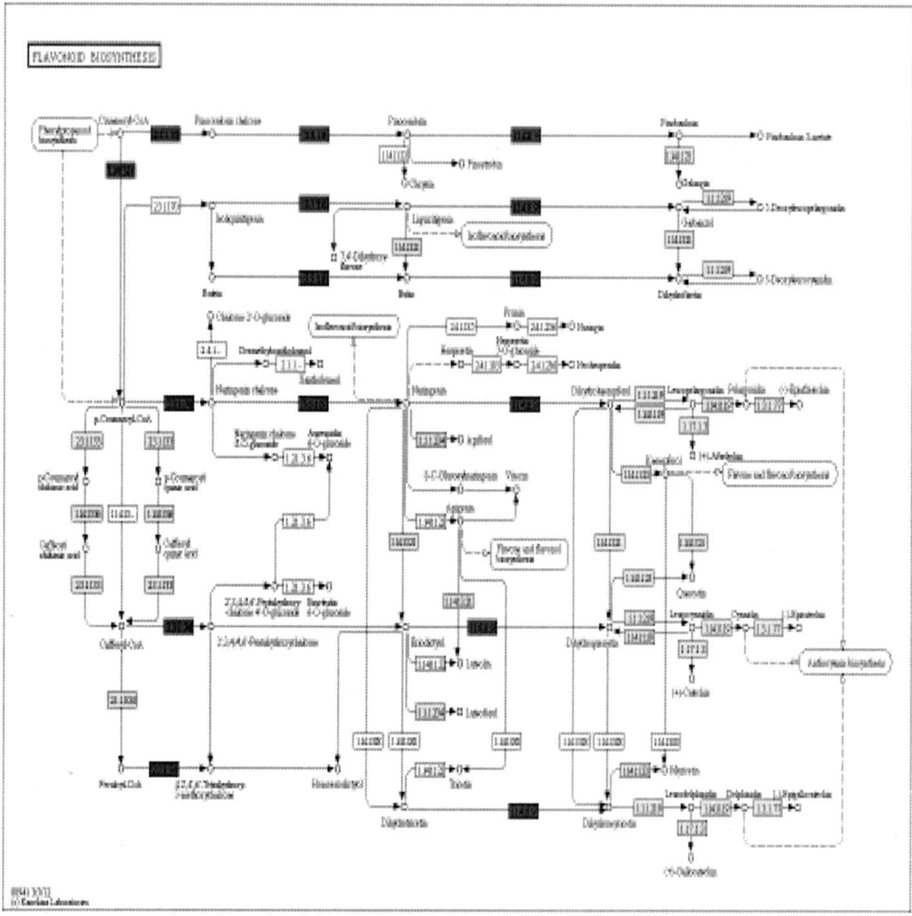

Figure S1.

Schematic representation of the flavonoid biosynthesis pathway. Each box represents a structural gene encoding a key enzyme involved in the flavonoid biosynthesis pathway. Numbers in each box are EC codes of each gene. Genes in red and green boxes represent those captured by our sequence data, with red boxes indicating genes expressed significantly higher in red than in white petals, and green boxes corresponding to genes with insignificant expression differences between colors. Uncolored boxes indicate uncaptured genes. EC code definitions can be found at: http://www.genome.jp/kegg-bin/show_pathway?map00941.

doi:10.1371/journal.pone.0090842.s001

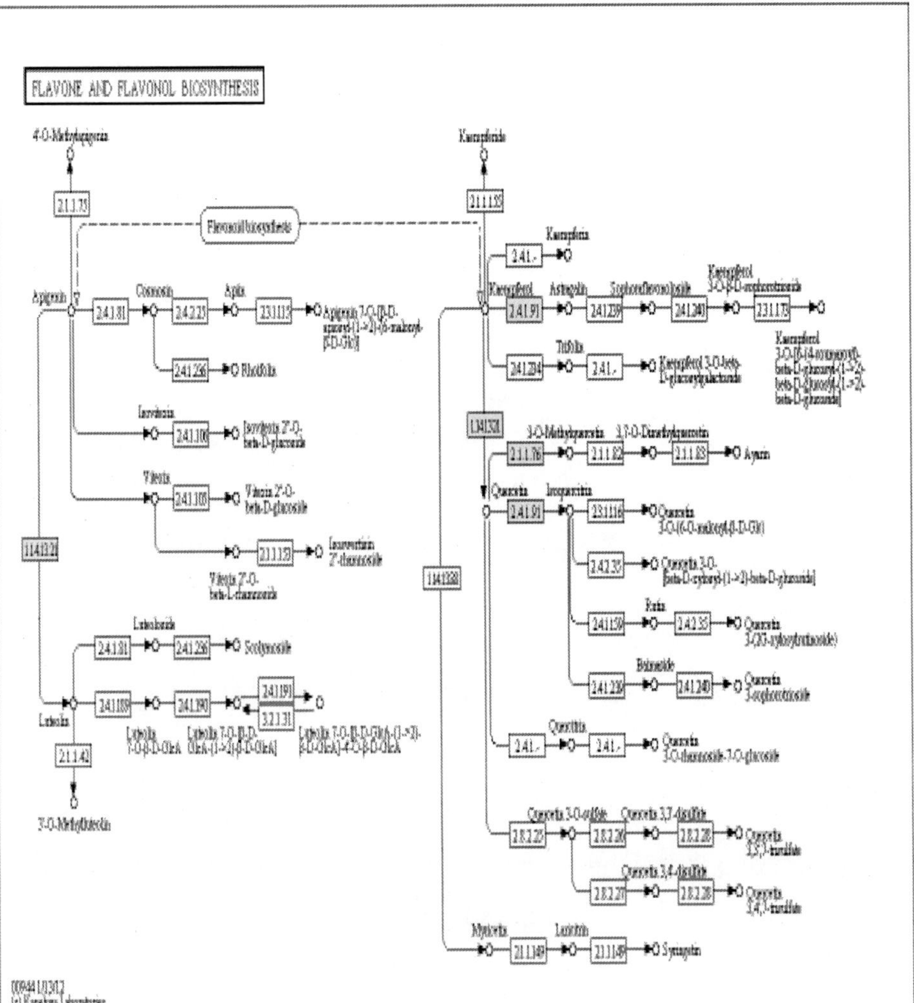

Figure S2.

Schematic representation of the flavone and flavonol biosynthesis pathway. Each box represents a structural gene encoding a key enzyme involved in the flavone and flavonol biosynthesis pathway. Numbers in each box are EC codes of each gene. Genes in green boxes represent those captured by our sequence data with insignificant expression differences between colors. Uncolored boxes correspond to uncaptured genes. EC code definitions can be found at: http://www.genome.jp/kegg-bin/show_pathway?map00944.

doi:10.1371/journal.pone.0090842.s002

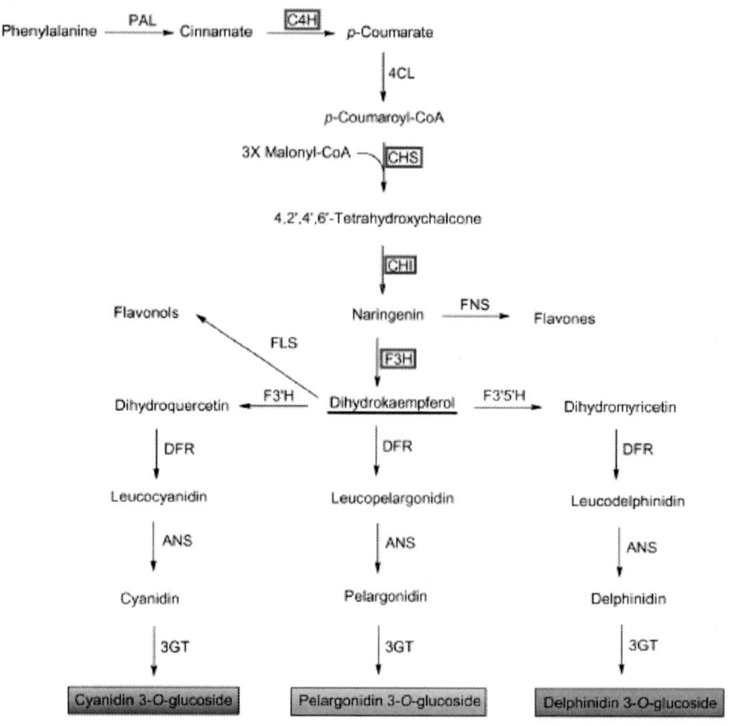

Figure S3.

Mapping of enzymes coded by differentially expressed structural genes to the flavonoid biosynthetic pathway. Enzymes corresponding to each of the differentially expressed structural genes are outlined in red.

doi:10.1371/journal.pone.0090842.s003

Figure S4.

Natural occurrence of different flower colors in peach (Prunus persica f. versicolor [Sieb.] Voss). (A) Flowers having different colors on the same branch. (B) A chimeric flower composed of white and red sections.

doi:10.1371/journal.pone.0090842.s004

AUTHOR CONTRIBUTIONS

Conceived and designed the experiments: TMY. Performed the experiments: YM HLL. Analyzed the data: YNC YM SXL FXY. Contributed reagents/materials/analysis tools: YM HLL SXL. Wrote the paper: YNC TMY YM FXY SXL.

REFERENCES

1. Kevan PG, Baker HG (1983) Insects as flower visitors and pollinators. Annu Rev Entomol 28: 407–53. View Article PubMed/NCBI Google Scholar
2. Harborne JB, Williams CA (2000) Advances in flavonoid research since 1992. Phytochemistry 55: 481–504. View Article PubMed/NCBI Google Scholar

3. Grotewold EA (2006) The genetics and biochemistry of floral pigments Rev. Plant Biol 57: 761–80. View Article PubMed/NCBI Google Scholar
4. Goto T, Kondo T (1991) Structure and molecular stacking of anthocyanins–flower color variation. Angew Chem Int Ed Engl 30: 17–33. View Article PubMed/NCBI Google Scholar
5. Holton TA, Cornish EC (1995) Genetics and biochemistry of anthocyanin biosynthesis. Plant Cell 7: 1071–1083. View Article PubMed/NCBI Google Scholar
6. Winkel-Shirley B (2001) Flavonoid biosynthesis: a colorful model for genetics, biochemistry, cell biology, and biotechnology. Plant Physiol 126: 485–493. View Article PubMed/NCBI Google Scholar
7. Davies KM (2009) Modifying anthocyanin production in flowers. In Anthocyanins. Edited by Winefield C, Davies K, Gould K. New York, Springer, 49–80.
8. Kroon J, Souer E, de Graaff A, Xue Y, Mol J, et al. (1994) Cloning and structural analysis of the anthocyanin pigmentation locus Rt of Petunia hybrida: characterization of insertion sequences in two mutant alleles. Plant J 5: 69–80. View Article PubMed/NCBI Google Scholar
9. Ronchi A, Petroni K, Tonelli C (1995) The reduced expression of endogenous duplications (REED) in the maize R gene family is mediated by DNA methylation. EMBO J 14: 5318–5328. View Article PubMed/NCBI Google Scholar
10. Fujiwara H, Tanaka Y, Fukui Y, Nakao M, Ashikari T, et al. (1997) Anthocyanin 5-aromatic acyltransferase from Gentiana triflora: purification, characterization and its role in anthocyanin biosynthesis. Eur J Biochem 249: 45–51. View Article PubMed/NCBI Google Scholar
11. Yoshida K, Toyama Y, Kameda K, Kondo T (2000) Contribution of each caffeoyl residue of the pigment molecule of gentiodelphin to blue color development. Phytochemistry 54: 85–92. View Article PubMed/NCBI Google Scholar
12. Yabuya T, Yamaguchi M, Fukui Y, Katoh K, Imayama T, et al. (2001) Characterization of anthocyanin p-coumaroyltransferase in flowers of Iris ensata. Plant Sci 160: 499–503. View Article PubMed/NCBI Google Scholar
13. Bruce W, Folkerts O, Garnaat C, Crasta O, Roth B, et al. (2000) Expression profiling of the maize flavonoid pathway genes controlled by estradiol-inducible transcription factors CRC and P. Plant Cell. 12: 65–80. View Article PubMed/NCBI Google Scholar
14. Morohashi K, Casas MI, Falcone Ferreyra ML, Mejía-Guerra MK,

Pourcel L, et al. (2012) A genome-wide regulatory framework identifies maize pericarp color1 controlled genes. Plant Cell 24: 2745–2764. View Article PubMed/NCBI Google Scholar

15. Mol J, Grotewold E (1998) KoesR (1998) How genes paint flowers and seeds. Trends Plant Sci 3: 212–217 View Article PubMed/NCBI Google Scholar
16. Forkmann G, de Vlaming P, Spribille R, Wiering H, Schram AW (1986) Genetic and biochemical studies on the conversion of dihydroflavonols to flavonols and flavones of Petunia hybrida. Z Naturforsch C: Biosci 41c: 179–186. View Article PubMed/NCBI Google Scholar
17. Dixon RA, Lamb CJ, Masoud S, Sewalt VJ, Paiva NL (1996) Metabolic engineering: prospects for crop improvement through the genetic manipulation of phenylpropanoid biosynthesis and defense responses-a review. Gene 179: 61–71. View Article PubMed/NCBI Google Scholar
18. Suzuki S, Nishihara M, Nakatsuka T, Misawa N, Ogiwara I, et al. (2007) Flower color alteration in Lotus japonicus by modification of the carotenoid biosynthetic pathway. Plant Cell Rep 26: 951–959. View Article PubMed/NCBI Google Scholar
19. Davies KM, Albert NW, Schwinn KE (2012) From landing lights to mimicry: the molecular regulation of flower colouration and mechanisms for pigmentation patterning. Funct Plant Biol 39: 619–638. View Article PubMed/NCBI Google Scholar
20. Dubois A, Carrere S, Raymond O, Pouvreau B, Cottret L, et al. (2012) Transcriptome database resource and gene expression atlas for the rose. BMC Genomics 13: 638. View Article PubMed/NCBI Google Scholar
21. To KY, Wang CK (2006) Molecular breeding of flower color. In Floriculture, Ornamental and Plant Biotechnology: Advances and Topical Issues Volume I. Edited by Teixeira da Silva JA. Isleworth, England, 300–310.
22. Iida S, Hirota T, Morisaki T, Marumoto T, Hara T, et al. (2004) Tumor suppressor WARTS ensures genomic integrity by regulating both mitotic progression and G1 tetraploidy checkpoint function. Oncogene 23: 5266–5274. View Article PubMed/NCBI Google Scholar
23. van Houwelingen A, Souer E, Spelt K, Kloos D, Mol J, et al. (1998) Analysis of flower pigmentation mutants generated by random transposon mutagenesis in Petunia hybrida. Plant J 13: 39–50. View Article PubMed/NCBI Google Scholar
24. Iida S, Hoshino A, Johzuka-Hisatomi Y, Habu Y, Inagaki Y (1999) Floricultural traits and transposable elements in the Japanese and

common morning glories. Ann NY Acad Sci 18: 265–74. View Article PubMed/NCBI Google Scholar

25. Liu D, Galli M, Crawford NM (2001) Engineering variegated floral patterns in tobacco plants using the Arabidopsis transposable element Tag1. Plant Cell Physiol 42: 419–423. View Article PubMed/NCBI Google Scholar
26. Inagaki Y, Hisatomi Y, Suzuki T, Kasahara K, Iida S (1994) Isolation of a Suppressor-mutator/Enhancer-like, transposable element, Tpn1, from Japanese morning glory bearing variegated flowers. Plant Cell 6: 375–383. View Article PubMed/NCBI Google Scholar
27. Hoshino A, Inagaki Y, Iida S (1995) Structural analysis of Tpnl, a transposable element from Japanese morning glory bearing variegated flowers. Mol Gen Genet 247: 114–117. View Article PubMed/NCBI Google Scholar
28. Quattrocchio F, Wing J, van der Woude K, Souer E, deVetten N, et al. (1999) Molecular analysis of the anthocyanin2 gene of petunia and its role in the evolution of flower color. Plant Cell 11: 1433–1444. View Article PubMed/NCBI Google Scholar
29. Chaparro JX, Werner D, Whetten R, O'Malley D (1995) Characterization of an unstable anthocyanin phenotype and estimation of somatic mutation rates in peach. J. Hered 86: 186–193. View Article PubMed/NCBI Google Scholar
30. Verde I, Abbott AG, Scalabrin S, Jung S, Shu SQ, et al. (2013) The high-quality draft genome of peach (Prunus persica) identifies unique patterns of genetic diversity, domestication and genome evolution. Nat Genet 45: 487–494. View Article PubMed/NCBI Google Scholar
31. Wang L, Zhao S, Gu C, Zhou Y, Ma JJ, et al. (2013) Deep RNA-Seq uncovers the peach transcriptome landscape. Plant Mol Biol 83: 365–377. View Article PubMed/NCBI Google Scholar
32. Mardis ER (2008) The impact of next-generation sequencing technology on genetics. Trends Genet 24: 133–141. View Article PubMed/NCBI Google Scholar
33. Morozova O, Marra MA (2008) Applications of next-generation sequencing technologies in functional genomics. Genomics 92: 255–264. View Article PubMed/NCBI Google Scholar
34. Parchman TL, Geist KS, Grahnen JA, Benkman CW, Buerkle CA (2010) Transcriptome sequencing in an ecologically important tree species: assembly, annotation, and marker discovery. BMC Genomics 11: 180. View Article PubMed/NCBI Google Scholar
35. Zhang X, Byrnes JK, Gal TS, Li WH, Borevitz JO (2008) Whole genome

transcriptome polymorphisms in Arabidopsis thaliana. Genome Biol 9: R165. View Article PubMed/NCBI Google Scholar

36. Wu B, Li Y, Yan HX, Ma YM, Luo HM, et al. (2012) Comprehensive transcriptome analysis reveals novel genes involved in cardiac glycoside biosynthesis and mlncRNAs associated with secondary metabolism and stress response in Digitalis purpurea. BMC Genomics 13: 15. View Article PubMed/NCBI Google Scholar

37. Huang LL, Yang X, Sun P, Tong W, Hu SQ (2012) The first illumina-based de novo transcriptome sequencing and analysis of safflower flowers. PLoS ONE 7: e38653. doi 0.1371/journal.pone.0038653.

38. Mahomed W, van den Berg N (2011) EST sequencing and gene expression profiling of defence-related genes from Persea americana infected with Phytophthora cinnamomi. BMC Plant Biol 11: 167. View Article PubMed/NCBI Google Scholar

39. Liu J, Yin T, Ye N, Chen Y, Yin T, et al. (2013) Transcriptome analysis of the differentially expressed Genes in the Male and Female Shrub Willows (Salix suchowensis). PLoS ONE 8: e60181 doi:10.1371/journal.pone.0060181. View Article PubMed/NCBI Google Scholar

40. Quinlan RJ, Tobin JL, Beales PL (2008) Chapter 5 modeling ciliopathies: primary cilia in development and disease. Curr Top Dev Biol 84: 249–310. View Article PubMed/NCBI Google Scholar

41. 41. Altschul SF, Gish W, Miller W, Meyers EW, Lipman DJ (1990) Basic local alignment search tool. J Mol Biol 215: 403–410. View Article PubMed/NCBI Google Scholar

42. Altschul SF, Madden TL, Schaffer AA, Zhang J, Zhang Z, et al. (1997) Gapped BLAST and PSI-BLAST: a new generation of protein database search programs. Nucleic Acids Res 25: 3389–3402. View Article PubMed/NCBI Google Scholar

43. The UniProt Consortium (2008) The Universal Protein Resource (UniProt). Nucleic Acids Res 36: D190–D195. View Article PubMed/NCBI Google Scholar

44. Aoki-Kinoshita KF, Kanehisa M (2007) Gene annotation and pathway mapping in KEGG. Methods Mol Biol 396: 71–91. View Article PubMed/NCBI Google Scholar

45. Moriya Y, Itoh M, Okuda S, Yoshizawa AC, Kanehisa M (2007) KAAS: an automatic genome annotation and pathway reconstruction server. Nucleic Acids Res 35: W182–W185. View Article PubMed/NCBI Google Scholar

46. Stekel DJ, Git Y, Falciani F (2000) The comparison of gene expression from multiple cDNA Libraries. Genome Res 10: 2055–2061. View Article PubMed/NCBI Google Scholar

47. Romualdi C, Bortoluzzi S, D'Alessi F, Danieli GA (2003) IDEG6: a web tool for detection of differentially expressed genes in multiple tag sampling experiments. Physiol Genomics 12: 159–62. View Article PubMed/NCBI Google Scholar
48. Livak KJ, Schmittgen TD (2001) Analysis of Relative Gene Expression Data Using Real-Time Quantitative PCR and the 2–ΔΔCT Method. Methods 25: 402–408. View Article PubMed/NCBI Google Scholar
49. Tong ZG, Gao ZH, Wang F, Zhou J, Zhang Z (2009) Selection of reliable reference genes for gene expression studies in peach using real-time PCR. BMC Mol Biol 10: 71. View Article PubMed/NCBI Google Scholar
50. Zhang Z, Wang Y, Wang S, Liu J, Warren W, et al. (2011) Transcriptome analysis of female and male Xiphophorus maculatus Jp 163 A. PLoS ONE. 6: e18379 doi:10.1371/journal.pone.0018379. View Article PubMed/NCBI Google Scholar
51. Broun P (2005) Transcriptional control of flavonoid biosynthesis: a complex network of conserved regulators involved in multiple aspects of differentiation in multiple aspects of in differentiation Arabidopsis. Curr Opin Plant Biol 8: 272–279. View Article PubMed/NCBI Google Scholar
52. Morita Y, Saitoh M, Hoshino A, Nitasaka E, Iida S (2006) Isolation of cDNAs for R2R3-MYB, bHLH and WDR transcriptional regulators and identification of c and ca mutations conferring white flowers in the Japanese morning glory. Plant Cell Physiol 47: 457–470. View Article PubMed/NCBI Google Scholar
53. Schmid M, Davison TS, Henz SR, Pape UJ, Demar M, et al. (2005) A gene expression map of Arabidopsis thaliana development. Nat Genet 37: 501–506. View Article PubMed/NCBI Google Scholar
54. Ramskold D, Wang ET, Burge CB, Sandberg R (2009) An abundance of ubiquitously expressed genes revealed by tissue transcriptome sequence data. PLoS Comput Biol 5: e1000598 doi:10.1371/journal.pcbi.1000598. View Article PubMed/NCBI Google Scholar
55. Guo SG, Zheng Y, Joung JG, Liu SQ, Zhang ZH, et al. (2010) Transcriptome sequencing and comparative analysis of cucumber flowers with different sex types. BMC Genomics 11: 384. View Article PubMed/NCBI Google Scholar
56. Zhu ML, Zheng XC, Shu QY, Li H, Zhong PX, et al. (2012) Relationship between the composition of flavonoids and flower colors variation in tropical water lily (Nymphaea) cultivars. PloS ONE 7: e34335 doi:10.1371/journal.pone.0034335. View Article PubMed/NCBI Google Scholar

57. van der Krol AR, Lenting PE, Veenstra J, van der Meer IM, Koes RE, et al. (1988) An antisense chalcone synthase gene in transgenic plants inhibits flower pigmentation. Nature 333: 866–869. View Article PubMed/NCBI Google Scholar
58. Napoli C, Lemieux C, Jorgensen R (1990) Introduction of a chimeric chalcone synthase gene into petunia results in reversible co-suppression of homologous genes in trans. Plant Cell 2: 279–289. View Article PubMed/NCBI Google Scholar
59. Gutterson N (1995) Anthocyanin biosynthetic genes and their application to flower color modification through sense suppression. Hort Sci 30: 964–966. View Article PubMed/NCBI Google Scholar
60. Zuker A, Tzfira T, Ben-Meir H, Ovadis M, Shklarman E, et al. (2002) Modification of flower color and fragrance by antisense suppression of the flavanone 3-hydroxylase gene. Mol Breeding 9: 33–41. View Article PubMed/NCBI Google Scholar
61. Schwinn K, Venail J, Shang Y, Mackay S, Alm V, et al. (2006) A small family of MYB-regulatory genes controls floral pigmentation intensity and patterning in the genus Antirrhinum. Plant Cell 18: 831–851.

Chapter 6

HIGH-DENSITY SNP GENOTYPING OF TOMATO (SOLANUM LYCOPERSICUM L.) REVEALS PATTERNS OF GENETIC VARIATION DUE TO BREEDING

[1]Sung-Chur Sim, [2]Allen Van Deynze, [2]Kevin Stoffel, [3]David S. Douches, [3]Daniel Zarka, [4]Martin W. Ganal, [5]Roger T. Chetelat, [6]Samuel F. Hutton, [6]John W. Scott, [7]Randolph G. Gardner, [7]Dilip R. Panthee, [8]Martha Mutschler, [9]James R. Myers,[1]David M. Francis

Department of Horticulture and Crop Science, The Ohio State University, Ohio Agricultural Research and Development Center, Wooster, Ohio, United States of America

[2]Seed Biotechnology Center, University of California Davis, Davis, California, United States of America

[3]Department of Crop and Soil Sciences, Michigan State University, East Lansing, Michigan, United States of America

[4]TraitGenetics GmbH, Gatersleben, Germany

[5]C. M. Rick Tomato Genetic Resource Center, University of California Davis, Davis, California, United States of America

[6]University of Florida, Gulf Coast Research and Education Center, Wimauma, Florida, United States of America

[7]Department of Horticultural Science, North Carolina State University, Mountain Horticultural Crops Research and Extension Center, Mills River, North Carolina, United States of America

[8]Department of Plant Breeding and Genetics, Cornell University, Ithaca, New York, United States of America

[9]Department of Horticulture, Oregon State University, Corvallis, Oregon, United States of America

ABSTRACT

The effects of selection on genome variation were investigated and visualized in tomato using a high-density single nucleotide polymorphism (SNP) array. 7,720 SNPs were genotyped on a collection of 426 tomato accessions (410 inbreds and 16 hybrids) and over 97% of the markers were polymorphic in the entire collection. Principal component analysis (PCA) and pairwise estimates of Fst supported that the inbred accessions represented seven sub-populations including processing, large-fruited fresh market, large-fruited vintage, cultivated cherry, landrace, wild cherry, and S. pimpinellifolium. Further divisions were found within both the contemporary processing and fresh market sub-populations. These sub-populations showed higher levels of genetic diversity relative to the vintage sub-population. The array provided a large number of polymorphic SNP markers across each sub-population, ranging from 3,159 in the vintage accessions to 6,234 in the cultivated cherry accessions. Visualization of minor allele frequency revealed regions of the genome that distinguished three representative sub-populations of cultivated tomato (processing, fresh market, and vintage), particularly on chromosomes 2, 4, 5, 6, and 11. The PCA loadings and Fst outlier analysis between these three sub-populations identified a large number of candidate loci under positive selection on chromosomes 4, 5, and 11. The extent of linkage disequilibrium (LD) was examined within each chromosome for these sub-populations. LD decay varied between chromosomes and sub-populations, with large differences reflective of breeding history. For example, on chromosome 11, decay occurred over 0.8 cM for processing accessions and over 19.7 cM for fresh market accessions. The observed SNP variation and LD decay suggest that different patterns of genetic variation in cultivated tomato are due to introgression from wild species and selection for market specialization.

INTRODUCTION

Selection of favorable alleles through domestication and breeding has led to dramatic changes in seed and fruit attributes, plant habit, and productivity. For tomato (Solanum lycopersicum L), breeding has involved the competing forces of narrowed genetic variation due to best by best crosses followed by selection [1], [2], and the expansion of genetic variation due to the introgression of genes for biotic stress resistance from wild species [3]–[5]. The long history of crossing to wild relatives has broadened the genetic diversity in contemporary germplasm relative to vintage and landrace germplasm [6]–[9]. In addition, breeding for distinct market classes and production systems has led to genetic differentiation in contemporary germplasm. For example, processing and field-grown fresh market tomatoes are now distinct sub-populations [8], [9].

Novel sequencing technologies have uncovered sufficient variation to investigate the effect of human selection across an entire plant genome [10]. Single nucleotide polymorphisms (SNPs) are a predominant form of sequence variation among individuals representing as much as 90% of the genetic variation in any species [11]. SNPs are distributed throughout a genome, they provide stable markers for genetic analysis, and their detection is amenable to automation. It is increasingly cost and time efficient to genotype large populations in a high-throughput manner. Because of these advantages, SNPs have become a marker system of choice for genetic analysis in plant species. In tomato, SNPs have been discovered using several methods: in silico mining of expressed sequence tag (EST) databases [12]–[14], intron and amplicon sequencing of conserved orthologous set (COS) genes [15], [16], and hybridization to oligonucleotide arrays [8]. Recently, 62,576 non-redundant SNPs were identified based on transcritpome sequences for six tomato accessions [10].

High-throughput SNP discovery has been accompanied by the development of array-based genotyping platforms that permit rapid scoring of several thousand markers in parallel [17]. Such SNP genotyping methods have facilitated high-density genetic map construction and genome-wide association analysis. For example, the use of a maize array with 49,585 SNPs produced two genetic linkage maps with 20,913 and 14,524 markers, respectively [18]. In tomato, we used the "SolCAP" array with 7,720 SNPs to generate high-density genetic maps using two F2 interspecific populations: 3,503 markers in the S. lycopersicum LA0925 x S. pennellii LA0714 (EXPEN 2000) population and 4,491 markers in the Moneymaker x S. pimpinellifolium LA0121 (EXPIM 2012) population [19]. An array with 44,100 SNPs was used to genotype 413 diverse accessions of rice and analyzed for association with 34 quantitative traits [20]. SNP data from such arrays can also be a resource for germplasm management in breeding programs and has a role in genomic selection strategies for crop improvement [18], [21].

The analysis of variation in tomato populations has often focused on differences between cultivated and wild species. In contrast, relatively little is known about which genes or genomic regions distinguish market classes within cultivated genepools. In the early 1900s, significant effort was placed on the evaluation of wild germplasm as a source of new resistance genes [3], [4]. Wide crosses were used to introduce new sources of resistance, with several varieties in commerce carrying introgressed genes by the late 1930s and early 1940s. These efforts often had a regional focus, and it remains unclear how introgressed regions are dispersed within breeding programs [22]. In addition, there was a concerted effort to breed for distinct plant habits and fruit characteristics for the processing and fresh market tomato industries. These efforts were initiated in the 1940s and resulted in the first cultivars suitable for machine harvest by the early 1960s [23]. At least three sources of S. pimpinellifolium were incorporated into processing breeding in order to develop compact plants amenable to "once over" (i.e. destructive) harvest [23].

In order to investigate genetic variation on the tomato genome due to contemporary breeding, we subjected a collection of 426 accessions to high-throughput genotyping using the SolCAP SNP array. The tomato accessions represented different market classes of cultivated tomato and closely related wild species. Knowledge of the genetic and physical organization of the SNPs [19] allowed us to conduct population level analysis based on the germplasm panel. SNP genotypes from the array were analyzed to identify sub-populations, and to assess genetic differentiation and diversity between and within sub-populations. We also investigated patterns of linkage disequilibrium (LD) within each chromosome in three sub-populations of large-fruited cultivated accessions (processing, fresh market, and vintage). The population level analysis revealed regions of the genome with high genetic variation between sub-populations, suggesting that historical breeding practices have led to different patterns of genetic variation in cultivated tomato germplasm.

RESULTS

Array-based SNP Genotyping

We used an array consisting of 7,720 SNPs distributed throughout the genome [10], [19] to genotype 426 accessions (410 inbreds and 16 hybrids; referred to as the SolCAP germplasm) (Table S1 and Table S2). The hybrids were chosen to maximize heterozygosity and develop a cluster file for the GenomeStudio software (Illumina Inc., San Diego, CA, USA). In order to establish an accurate and automatic genotype calling procedure for the GenomeStudio software that is usable across the entire genepool of cultivated tomato, clustering based on the SolCAP germplasm was cross-validated with a cluster file based on 92 hybrids developed independently by TraitGenetics [19]. The resulting high-quality cluster file is available through the eXtension website [24].

In order to assess the quality of SNP calls, we duplicated 34 accessions that were randomly selected using independent DNA preparations and genotyping facilities. The average proportion of consistent calls across all accessions was 98.7%. Cluster analysis was performed, and for all 34 accessions, the nearest neighbor was the duplicate accession. The proportion of consistent calls increased to 99.3% by excluding three accessions, OH981136 (processing, 91.8%), Purple Clabash (vintage, 90.0%), and Principle Borghese (vintage, 94.1%). One of the duplicate samples for OH981136 showed higher rates for no call (8.8%) and heterozygote call (8.4%) relative to the other sample (0.7% for no call and 0.2% for heterozygote call). The data quality between two samples of Purple Clabash was also different from each other (0.7% vs. 5.7% for no call). Heterozygote call rates were similar for the two samples. The duplicate samples of Principe Borghese differed by 5.9%, but showed similar rates for no call (0.8%

vs. 1.0%) and heterozygote call (0.2% vs. 0.2%). These results suggest that overall reproducibility is high, and that differential calls may result from DNA quality that affects the percentage of "no call", residual heterozygosity, and variation within accessions.

The data were analyzed to determine the polymorphism rate based on the inbred accessions. A total of 7,500 SNPs (97.2%) were polymorphic and 61 SNPs (0.8%) were monomorphic (Table 1). Among the 7,500 polymorphic SNPs, there were 7,375 SNPs with <10% missing data, 84 SNPs with 10–20% missing data, and 41 SNPs with >20% missing data. The SNPs with a high frequency of missing data were randomly distributed across accessions. Polymorphism of 123 SNPs could not be unequivocally determined because of a large amount (≥10.0%) of missing data (34 SNPs) or because polymorphism detection was due to a single homozygous allele and only heterozygote calls for the alternate allele (89 SNPs). It is possible that the heterozygotes actually represent duplicated genes, or the alternate allele was simply not present as a homozygote. These 89 SNPs were also randomly distributed across the genome. In addition, 36 SNPs failed to produce a genotype call because of poor signals (Table 1).

Table 1. Polymorphism of 7,720 SNP markers based on 410 inbred tomato accessions in the SolCAP germplasm collection.

doi:10.1371/journal.pone.0045520.t001

Class	No. of Marker	Percentage (%)
Polymorphic	7,500 (7,375[1])	97.2 (95.5)
Monomorphic	61	0.8
Undetermined[2]	123	1.6
No call	36	0.4
Total	7,720	100.00

[1]Number of SNPs with less than 10% missing data.
[2]The class includes SNPs that were either monomorphic with ≥10% missing data (34 SNPs) or polymorphic due to only heterozygote calls (89 SNPs).
doi:10.1371/journal.pone.0045520.t001

The proportion of heterozygous SNPs was assessed within cultivated tomato germplasm and S. pimpinellifolium accessions. The heterozygosity with all scorable markers (7,684 SNPs) was 0.02 for processing, 0.01 for large-fruited fresh market, 0.01 for large-fruited vintage, and 0.04 for S. pimpinellifolium accessions. Using only markers that were polymorphic within each sub-population (range 3,700– 6,022 polymorphic markers), the processing, fresh market, and vintage accessions showed heterozygosity levels of 0.02, 0.01, and 0.02, respectively. The heterozygosity for S. pimpinellifolium was 0.05.

Genetic Differentiation between Sub-populations

The 410 inbred accessions were first divided into five sub-populations based on a priori knowledge of germplasm pedigree, age, market class, and origin: 141 processing, 122 fresh market, 88 vintage, 43 wild cherry, and 16 S. pimpinellifolium. Principal component analysis (PCA) and linkage disequilibrium (LD) analysis supported separation of the fresh market sub-population into a group of 110 large-fruited accessions and 12 cherry accessions (cultivated cherry). Similarly, the vintage sub-population was divided into 61 large-fruited accessions, 15 cherry accessions which clustered with the cultivated cherry sub-population, and 12 landrace accessions (Figure 1 and Figure S1). This iterative analysis led to the definition of seven sub-populations for further analysis: 141 processing, 110 large-fruited fresh market (hereafter referred to as fresh market), 61 large-fruited vintage (hereafter referred to as vintage), 27 cultivated cherry, 12 landrace, 43 wild cherry, and 16 S. pimpinellifolium.

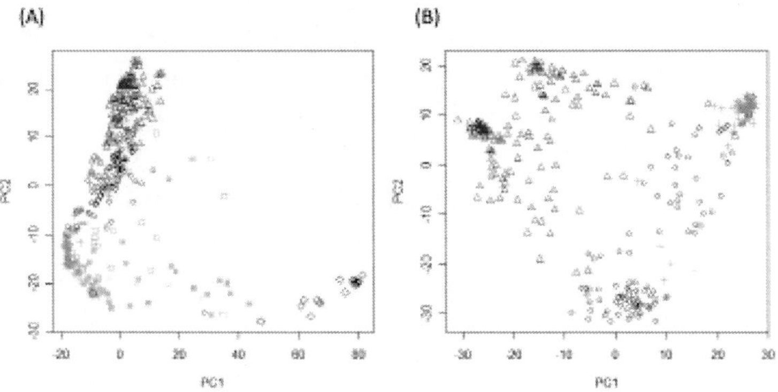

Figure 1. Principal component analysis (PCA) based on 4,393 SNP markers.

The PCA was conducted separately using data for all sub-populations of the SolCAP germplasm (A) and data for only the three large-fruited cultivated sub-populations consisting of the processing, fresh market, and vintage accessions (B). The processing accessions are indicated Δ (red); fresh market, ○ (blue); vintage, + (green); cultivated cherry, □ (violet); landrace, × (gold); wild cherry, ▪ (gray); and S. pimpinellifolium, ◊ (black).

doi:10.1371/journal.pone.0045520.g001

In the PCA analysis, the first two principal components (PC1 and PC2) explained 22% and 16% of the total variance, respectively. PC1 separated the S. pimpinellifolium accessions from all other sub-populations (P<0.001), while PC2 separated processing, fresh market, and vintage germplasm (P<0.01; Figure 1A). A subsequent PCA was conducted using only the processing, fresh market, and vintage sub-populations (Figure 1B). The second, more focused analysis validated the significant separation of the sub-populations. In addition, PCA based on only the cultivated accessions suggested further divisions within both processing and fresh market sub-populations (Figure 1B). Although most of the accessions were clustered into a corresponding sub-population, there were a few accessions that appear to be transitional and/or misclassified accessions. For example, Peto 460 (PI 600920; coordinates −22.1, −4.8 in Figure 1B) and Heinz 1370 (PI 341134; coordinates 4.6, −3.4 in Figure 1B) were grouped with the vintage accessions due to their date of release [25], but were developed as processing tomatoes. Similarly, Rio Grande (coordinates −12.5, 9.1 in Figure 1B) was classified as a fresh market accession, but clustered with the processing accession. This cultivar is an early "Roma" type tomato and was originally developed as a processing tomato. Clustering supports the processing origin of these three accessions (Figure 1B).

Pairwise analysis of Fst was used to test the significance of genetic differentiation between sub-populations (Table 2). In order to test effects of marker choice for this analysis, we estimated pairwise Fst using two independent subsets of SNP data derived from the array. Analysis was performed on all 410 inbred accessions for each marker set. In addition a re-sampling analysis was performed by randomly selecting n = 40 for processing, fresh market, and vintage sub-populations. Repetition of this analysis suggests that our conclusions are supported with balanced populations and across multiple marker subsets. Cultivated germplasm, including processing, fresh market, and vintage accessions were all significantly diverged (Fst = 0.29 to 0.41, P<0.005) (Table 2). The cultivated cherry, wild cherry, and landrace accessions were not differentiated from each other but were distinct from all other sub-populations (Fst = 0.13 to 0.64, P<0.005). The S. pimpinellifolium accessions were separated from all other sub-populations with estimates of pairwise Fst ranging between 0.57–0.81 (P<0.005; Table 2). The Fst analysis verified genetic differentiation between sub-populations defined by PCA.

Table 2. Pairwise estimates of Fst (θ) between sub-populations.

doi:10.1371/journal.pone.0045520.t002

Sub-population	Proc		FM		Vintage	Cherry	Landrace	Wild cherry	Pimp
Processing (Proc)	0.00		0.29**		0.41**	0.27**	0.40**	0.38**	0.72**
Fresh market (FM)			0.00		0.27**	0.18**	0.28**	0.29**	0.72**
Vintage					0.00	0.13**	0.18**	0.20**	0.81**
Cultivated cherry (Cherry)						0.00	0.04^{NS}	0.05*	0.58**
Landrace							0.00	0.04^{NS}	0.64**
Wild cherry								0.00	0.57**
S. pimpinellifolium (Pimp)									0.00
Further division within sub-population	Proc 1	Proc 2	FM 1	FM 2	Vintage	Cherry	Landrace	Wild cherry	Pimp
Proc 1	0.00	0.27**	0.42**	0.40**	0.52**	0.34**	0.49**	0.44**	0.74**
Proc 2		0.00	0.52**	0.32**	0.47**	0.30**	0.45**	0.38**	0.75**
FM 1			0.00	0.32**	0.52**	0.34**	0.49**	0.42**	0.76**
FM 2				0.00	0.12**	0.10**	0.17**	0.19**	0.72**

Pairwise θ (92) was estimated using the Microsatellite analyzer v4.05 (93). P-value was calculated based on 10,000 permutations with Bonferroni correction. NS, not significant, *P<0.05 and **P<0.005.

doi:10.1371/journal.pone.0045520.t002

Further divisions within both processing and fresh market sub-populations detected in PCA were also tested by estimating pairwise Fst. The processing accessions can be divided into two groups consisting of 82 and 52 accessions, respectively (Table S3). Seven accessions were not grouped because they were outliers, these tended to be CULBPT accessions which derive from recent crosses to fresh-market material. The two main processing groups were significantly differentiated from each other (Fst = 0.27 P<0.005) and distinct from the other germplasm (Table 2). Two groups of the fresh market tomatoes including 61 and 49 accessions each showed significant differentiation between them and from the other sub-populations (Fst = 0.10 to 0.76, P<0.005; Table 2).

Levels of Polymorphism within Sub-populations

The level of genetic diversity within each sub-population was measured using allelic richness (A), expected heterozygosity (He), and polymorphic information content (PIC). Within cultivated germplasm, the highest estimates of A, He, and PIC were found in the cultivated cherry accessions (Table 3). Among the remaining cultivated sub-populations, the fresh market accessions showed higher A, while the landraces had the highest He and PIC values. The sub-population of vintage accessions contained the lowest variation for all three descriptors (Table 3). The further division of accessions within both processing and fresh market sub-populations showed higher estimates of these descriptive statistics relative to the vintage sub-population (Table 3).

Table 3. Descriptive statistics for genetic diversity within sub-populations.

doi:10.1371/journal.pone.0045520.t003

Sub-population[1]	Sample size	*A*[2]	*He*[3]	PIC[4]
Processing (Proc)	141	1.49	0.16	0.13
Proc 1	82	1.47	0.13	0.11
Proc 2	52	1.38	0.12	0.09
Fresh market (FM)	110	1.59	0.16	0.13
FM 1	61	1.46	0.11	0.09
FM 2	49	1.61	0.16	0.13
Vintage	61	1.37	0.09	0.07
Cultivated cherry	27	1.89	0.27	0.21
Landrace	12	1.54	0.19	0.14
Wild cherry	43	1.89	0.26	0.21
S. pimpinellifolium	16	1.87	0.28	0.21

[1]The further divisions within both processing and fresh market sub-populations are indicated with a number followed by each sub-population name (e.g. Proc 1 and Proc 2). Seven processing accessions were excluded for this grouping because they were outliners.
[2]Allelic richness [94,95].
[3]Expected heterozygosity corrected for sample size [96].
[4]Polymporphic Information Content [97].
doi:10.1371/journal.pone.0045520.t003

The highest number of polymorphic markers (6,234 SNPs) was identified in the cultivated cherry sub-population. There were

5,909 and 5,650 polymorphic markers in the wild cherry and S. pimpinellifolium sub-populations (Table 4). For the contemporary accessions, 4,648 and 6,022 markers were polymorphic in the processing and fresh market sub-populations, respectively (Table 4). Fewer polymorphic markers were found in the vintage (3,700 SNPs) and landrace (3,159 SNPs) sub-populations. Distribution patterns of polymorphic SNP markers across 12 chromosomes were different between sub-populations (Table 4). For example, chromosome 8 showed proportionally lower SNP numbers for the processing and cultivated cherry accessions, as did chromosome 12 for the large-fruited fresh market and wild cherry tomatoes. Chromosome 10 had lower polymorphism rates for the vintage and landrace groups compared to contemporary processing and large-fruited fresh market sub-populations.

Table 4. Distribution of polymorphic SNP markers in seven sub-populations of the SolCAP germplasm.

doi:10.1371/journal.pone.0045520.t004

	Processing		Fresh market		Vintage		Cultivated cherry		Landrace[1]		Wild cherry		*S. pimpinellifolium*	
Chr	SNP No.	%	SNP No.	%	SNP No.	%	SNP No.	%	SNP No.	%	SNP No.	%	SNP No.	%
1	266	5.7	345	5.7	210	5.7	369	5.9	232	7.3	446	7.5	370	6.5
2	456	9.8	764	12.7	336	9.1	664	10.7	352	11.1	763	12.9	689	12.2
3	450	9.7	497	8.3	321	8.7	543	8.7	340	10.8	564	9.5	486	8.6
4	707	15.2	751	12.5	574	15.5	789	12.7	422	13.4	672	11.4	674	11.9
5	666	14.3	634	10.5	520	14.1	684	11.0	207	6.6	499	8.4	617	10.9
6	436	9.4	602	10.0	215	5.8	638	10.2	248	7.9	417	7.1	406	7.2
7	159	3.4	307	5.1	188	5.1	305	4.9	228	7.2	338	5.7	353	6.2
8	147	3.2	292	4.8	146	3.9	255	4.1	198	6.3	326	5.5	280	5.0
9	209	4.5	403	6.7	163	4.4	419	6.7	164	5.2	348	5.9	306	5.4
10	159	3.4	332	5.5	120	3.2	337	5.4	126	4.0	363	6.1	312	5.5
11	785	16.9	791	13.1	740	20.0	904	14.5	395	12.5	864	14.6	869	15.4
12	203	4.4	290	4.8	160	4.3	317	5.1	241	7.6	298	5.0	279	4.9
unknown	5	0.1	14	0.2	7	0.2	10	0.2	6	0.2	11	0.2	9	0.2
Total	4,648	100.0	6,022	100.0	3,700	100.0	6,234	100.0	3,159	100.0	5,909	100.0	5,650	100.0

[1]Latin American Cultivar.
doi:10.1371/journal.pone.0045520.t004

The number of polymorphic markers in sub-populations increases with the size of the population. This fact led to concerns that our estimates of diversity might be skewed by population size differences despite the fact that A and He are adjusted for population size. To address the concern, we performed rarefaction analysis to estimate polymorphic marker accumulation curves [26]. For all seven sub-populations, the curves reached an asymptote within the population size sampled (Figure 2). The S. pimpinellifolium, wild cherry, and cultivated cherry sub-populations were the most diverse groups based on the curves. Within the large-fruited cultivated sub-populations, fresh-market accessions were more diverse than vintage accessions. The accumulation curves for the processing and vintage accessions merged at n = 60, despite the fact that they were well separated between n = 10 to 25. This result suggests that differences in number of polymorphisms detected between these sub-populations may be population size dependent.

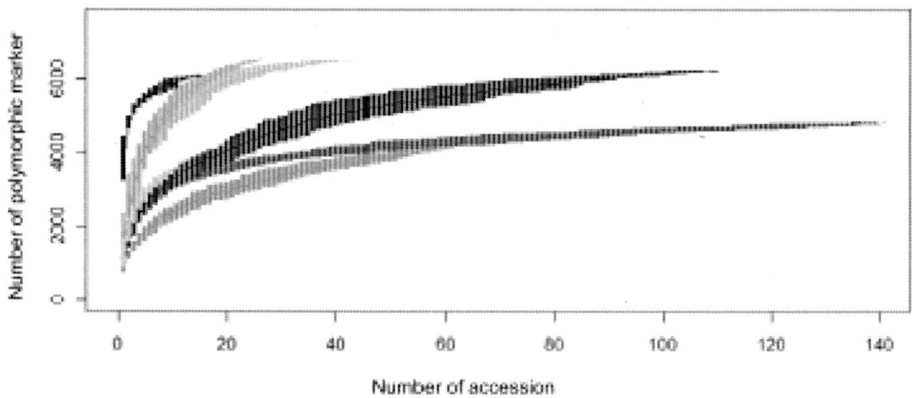

Figure 2. Rarefaction analysis to estimate the number of polymorphic markers in each sub-population.

Accessions are coded as in Figure 1 with processing indicated by red; fresh market by blue; vintage by green; cultivated cherry, violet; landrace, gold; wild cherry, gray; and S. pimpinellifolium, black. Curves are plotted with standard deviations indicated by vertical bars.

doi:10.1371/journal.pone.0045520.g002

Minor Allele Frequency of SNP Markers

Minor alleles for 7,310 SNP markers with physical map positions were determined based on genotypic data of 410 inbred accessions. Minor allele frequency (MAF) was then estimated within each sub-population (Table S4). We graphed MAF in order to visualize genetic variation between three representative sub-populations of cultivated tomatoes (processing, fresh market, and vintage) along with S. pimpinellifolium accessions. These plots revealed different MAF patterns over the entire genome with chromosomes 2, 4, 5, 6, and 11 being particularly variable between the cultivated sub-populations (Figure 3A and Figure 4A). The processing and fresh market sub-populations showed unique MAF patterns on these chromosomes relative to the vintage sub-population. MAF patterns on chromosome 5 distinguished the processing sub-population from the fresh market sub-population (Figure 3A and Figure 4A). Common alleles in the S. pimpinellifolium accessions tended to be minor alleles in the cultivated sub-populations (Figure 3A and Figure 4A). Since the processing and fresh market sub-populations were further divided into two sub-groups, respectively, MAF was estimated within each sub-group and graphed (Table S4 and Figure S2). The processing sub-groups were distinguished by unique MAF patterns on chromosomes 5 and 11. The most variable MAF patterns between the large-fruited fresh market sub-groups were found on chromosome 11 (Figure S2).

The minor allele was determined relative to allele calls for all 410 inbred accessions based on 7,310 SNPs, and then MAF was estimated and graphed for processing (Proc), fresh market (FM), and vintage (Vint) and S. pimpinellifolium (Pimp) sub-populations (A). The 28 tomato genes listed in Table S9 are positioned based on their coding sequences (arrow) and flanking markers (dotted line). The Y-axis represents allele frequency and the X-axis represents physical positions of the SNPs oriented with respect to the tomato genome sequence [102]. PCA loadings for PC 1 and PC 2 are graphed with candidates for loci under positive selection based on Fst outlier analysis [27], [28] (B). The candidate loci are indicated by dots (•) with a color scheme indicating pairwise comparisons that were significant: red for Proc vs. FM; green for Vint vs. Proc; blue for Vint vs. FM; violet for Proc vs. FM and Vint; sky blue for FM vs. Proc and Vint; and gold for Vint vs. Proc and FM. All other loci are indicated by X (black).

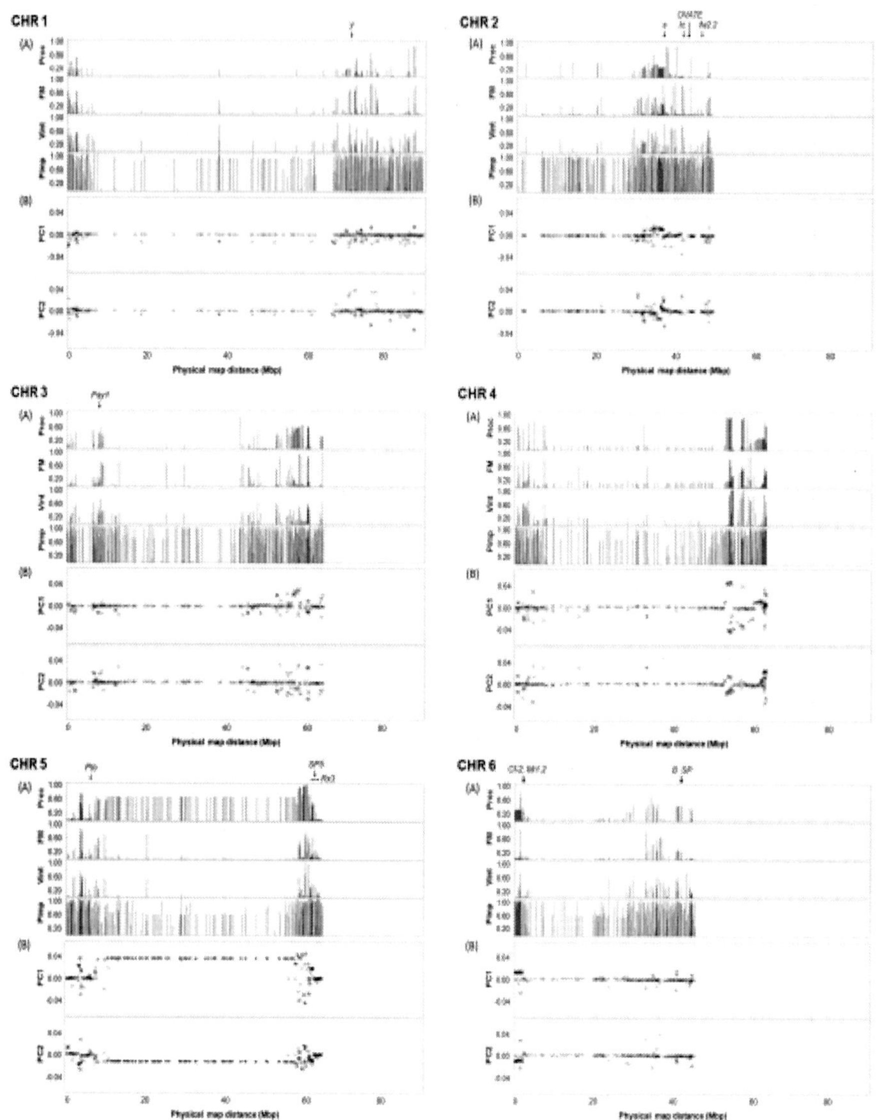

Figure 3. Minor allele frequency (MAF) patterns, PCA loadings, and Fst outliers on chromosomes 1 to 6.

doi:10.1371/journal.pone.0045520.g003

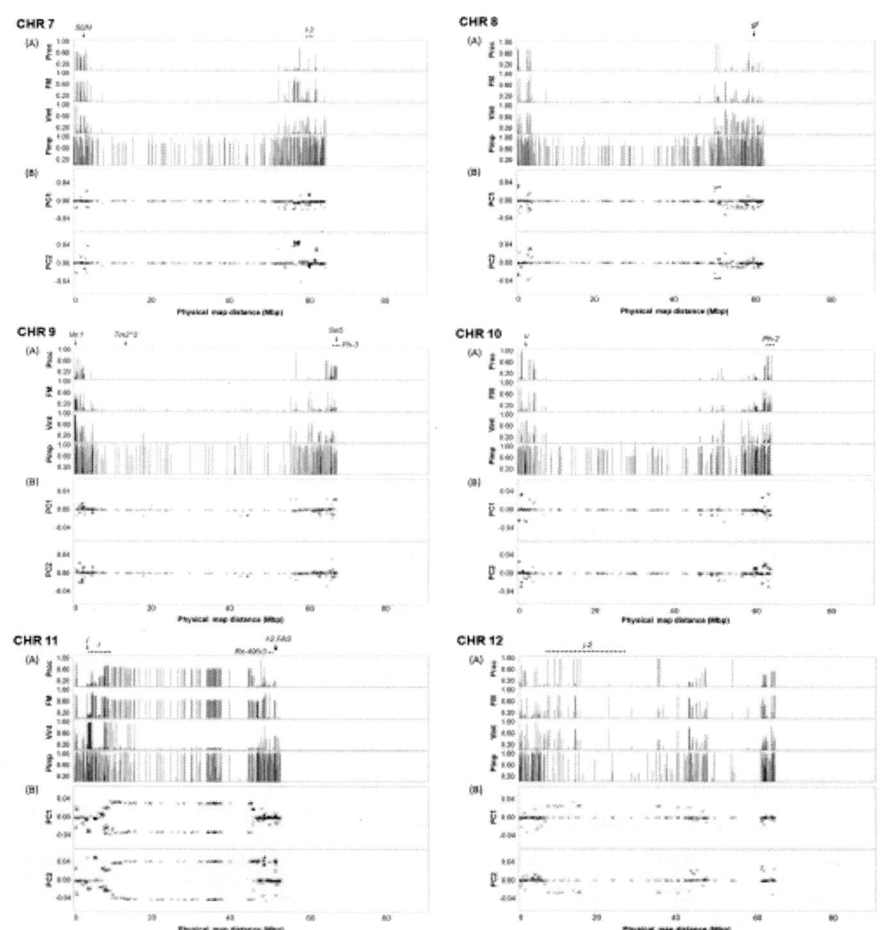

Figure 4. Minor allele frequency (MAF) patterns, PCA loadings, and Fst outliers on chromosomes 7 to 12.

The minor allele was determined relative to allele calls for all 410 inbred accessions based on 7,310 SNPs, and then MAF was estimated and graphed for processing (Proc), fresh market (FM), and vintage (Vint) and S. pimpinellifolium (Pimp) sub-populations (A). The 28 tomato genes listed in Table S9 are positioned based on their coding sequences (arrow) and flanking markers (dotted line). The Y-axis represents allele frequency and the X-axis represents physical positions of the SNPs oriented with respect to the tomato genome sequence [102]. PCA loadings for PC 1 and PC 2 are graphed with candidates for loci under positive selection based on Fst outlier analysis [27], [28] (B). The candidate loci are indicated by dots (•) with a color scheme indicating pairwise comparisons that were significant: red for Proc vs. FM; green for Vint vs. Proc; blue for Vint vs. FM; violet for Proc vs. FM and Vint; sky blue for FM vs. Proc and Vint; and gold for Vint vs. Proc and FM. All other loci are indicated by X (black).

doi:10.1371/journal.pone.0045520.g004

Loci Explaining Variation within Germplasm

PCA loadings measure the correlation between PC and SNP, and provide an estimate of how much each SNP contributes to variance. We extracted loadings from the PCA of only cultivated germplasm (Figure 1B) and displayed these values relative to physical position in order to visualize SNPs that contribute most to variance in the germplasm panel. There were 778 SNPs with absolute values of >0.02 for PC1. Most of these SNPs were found on chromosomes 4 (24.0%), 5 (32.6%), and 11 (31.1%). For PC2, 709 SNPs with absolute values of >0.02 were found on chromosomes 5 (12.3%) and 11 (52.9%) (Figure 3B, Figure 4B, and Table S5).

We also investigated loci that may be under positive selection between the three cultivated sub-populations using an Fst outlier method based the expected distribution of Fst and He [27], [28]. We identified 339 candidates for loci under positive selection between processing and fresh market as falling outside of the 95% confidence interval (Figure 3B, Figure 4B, and Table S6). A high portion of these loci (61.4%) were derived from chromosome 5, while 0.6–10.0% of the SNPs were distributed across the other 11 chromosomes. Comparison between processing and vintage germplasm detected 128 candidates for loci under positive selection (Figure 3B, Figure 4B, and Table S7). Among these, 57 loci (44.5%) were located on chromosome 4 and 35 loci (27.3%) on chromosome 5. For comparison between fresh market and vintage sub-germplasm, 208 loci were outliers based on a 95% confidence interval (Figure 3B, Figure 4B, and Table S8). Most of the loci were derived from chromosome 4 (42.6%) and chromosome 11 (43.5%). For all three pairwise comparisons, the candidate loci under positive selection were not randomly distributed within a chromosome.

To visualize the position of candidate genes that may have been selected for during breeding, we superimposed the position of 28 loci onto the physical map of MAF pattern, PCA loading, and Fst outlier detection. Candidate loci included genes that affect fruit size, shape, and color, disease resistance, and plant morphology (Table S9). Three genes for fruit shape and size, fw2.2, OVATE, and lc are located on chromosome 2 [29]–[32]. Of these genes, only lc appears to be a candidate for a locus under selection based on MAF patterns, PCA loadings, and Fst outlier detection of linked SNPs (Figure 3). The OVATE locus is polymorphic within all cultivated sub-populations, and therefore SNPs lack power to discriminate populations. The large fruited allele of fw2.2 is fixed in the cultivated sub-populations. The SUN mutation which controls elongated fruit shape is found on chromosome 7 [33] and is present at a low frequency in both processing and vintage accessions (Figure 4). The genomic region for the SUN mutation contains some SNPs with high absolute values for PCA loadings, though none of these were detected in the Fst outlier analysis. Another fruit shape gene, FAS is found in a region of chromosome 11 [34] with high loadings for PC2, and LD between FAS and SNPs in this region may be responsible for detection of Fst outliers between processing (which lack FAS) and vintage (which are polymorphic for FAS) germplasm (Figure 4). Phenotypic data are available in flat file format through the SolCAP website [35] and in searchable form through the Sol Genome Network (SGN) ontology database using advanced search options with the stock number prefix, SCT; stock type, accession; stock editors, SolCAP project; and the organism as either Solanum lycopersicum or S. pimpinellifolium [36]. A detailed analysis of phenotypic data is provided in a separate publication [37].

Fruit color genes affecting skin and flesh color are distributed on chromosomes 1 (y), 3 (Psy1), 6 (B), 8 (gf), and 10 (u) [38]–[43]. Although y, Psy1 and gf are found in regions of the genome with SNPs that have moderately high absolute values associated with PCA loadings and somewhat variable MAF between sub-populations, these regions do not appear to be highly discriminatory (Figure 3 and Figure 4). Regions on chromosome 1 and 3 were also not coincident with loci identified by the Fst outlier approach. The old gold crimson (ogc) allele of the B gene encodes a mutation in the fruit specific lycopene beta-cyclase [41] and is segregating in all three cultivated populations.

Thus minor variation in MAF patterns, PCA loadings, and the detection of a SNP under selection between processing and vintage are more likely a reflection of the closely linked SP allele. In contrast, the region of chromosome 8 containing gf contains several SNPs detected as outliers based on Fst between processing and vintage germplasm (Figure 4), which is consistent with the absence of gf mutations in processing populations and the presence of the mutant alleles in several vintage accessions. The allele of u gene for uniform ripening appears to fall in a region that is fixed in the processing germplasm. This allele is present at a high frequency in fresh market germplasm and at a low frequency in vintage germplasm (Figure 4). The region contains several SNPs with high absolute values for PCA loadings and also SNPs detected as Fst outliers.

The signals of genetic differentiation on chromosomes 5, 6, and 11 might be due to allelic variation in disease resistance genes, and reflect different introgression histories. Two resistance genes to bacterial disease, Pto and Rx3 on chromosome 5 [44], [45] appears to be candidates for loci under selection based on MAF patterns, PCA loading, and Fst outlier detection (Figure 3). The region of chromosome 6 containing Cf-2 and Mi1.2 genes [46], [47] shows signals of selection distinguishing processing accessions from fresh market and vintage accessions (Figure 3).

In general, Cf genes are deployed more frequently in fresh market material while Mi has been widely used in processing germplasm in California, but rarely used in processing accessions in the Midwestern U.S. Chromosome 11 contains one of the oldest introgressions, I for Fusarium resistance [48], [49]. This gene falls in a region of the genome that distinguishes contemporary germplasm from vintage germplasm based on MAF patterns (Figure 4). The region also contains some SNPs with high PC loadings. SNPs were detected as outliers at the 95% confidence level between processing and vintage germplasm. The region of chromosome 11 containing Rx-4/Xv3 and I-2 [50], [51] appears to be discriminatory between the cultivated sub-populations based on MAF patterns, PC2 loadings, and outlier SNPs (Figure 4). Other regions with signals for loci under selection include chromosomes 7 (I-3) [48], [52], chromosome 9 (Ve1, Tm2∧2, Sw5, and Ph-3) [53]–[57], and chromosome 10 (Ph-2) [58] (Figure 4). The genomic regions for Ph-2 and Ph-3 show MAF patterns that distinguish processing from fresh market and vintage sub-populations (Figure 4). SNPs were also detected in the regions by PCA loadings and Fst outlier analysis. These resistance genes have been deployed in several processing tomatoes [59]. The region of the genome containing Tm2∧2 distinguishes fresh-market from vintage and processing accessions based on the presence of SNPs with a MAF of 10–15%, consistent with deployment of this resistance in fresh market accessions (Figure 4). However, this region was not detected based on PC loadings or Fst outlier detection, possibly due to low allele frequencies for markers in linkage disequilibrium with Tm2∧2.

Genes affecting plant morphology are distributed across several chromosomes. A region on chromosome 2 that appears to differentiate sub-populations contains the compound inflorescence gene (s) [60] (Figure 3). Chromosomes 5 and 6 contain genes (SP5 and SP, respectively) that control plant habit and may be key selection points in differentiating vintage from contemporary sub-germplasm [61] and processing from fresh market [62], [63]. The SP5 gene is in a region of the genome containing SNPs with high loadings for PC2, but does not appear to have been detected by Fst outlier analysis (Figure 3). The jointless gene (j) on chromosome 11 [64] falls in a region containing SNPs with high absolute values for PCA loadings and outlier SNPs (Figure 4). This region distinguishes processing from fresh market and vintage accessions, reflecting the near fixation of jointless accessions in this germplasm category. The genomic region for the j-2 gene on chromosome 12 [65] contains SNPs with high absolute values for PCA loadings (Figure 4).

Linkage Disequilibrium (LD) Analysis

The extent of LD across each chromosome was analyzed for the processing, large-fruited fresh market, and large-fruited vintage sub-populations. Pairwise r2 was calculated using 1,572 polymorphic SNP markers with MAF of ≥0.1 for processing, 1,504 for fresh market, and 700 for vintage accessions. The r2 values were plotted against the genetic distance, and curves of LD decay were fitted using locally weighted scatterplot smoothing (LOESS) [66] and non-linear regression (NLR) [67]. The LOESS and NLR methods estimated similar LD decay on 9 chromosomes for processing, 8 chromosomes for fresh market, and 11 chromosomes for vintage germplasm. In these chromosomes, the observed difference of the LD decay between the methods ranged from 0 to 4.9 cM (Table 5 and Figures 5, 6, 7). Over 10 cM difference for LD decay estimated by LOESS and NLR was found on chromosome 6 (28.1 cM for processing and 14.1 cM for fresh market); chromosome 8 (20.2 cM for processing); and chromosome 11 (9.6 cM for fresh market and >40 cM for vintage) (Table 5 and Figures 5, 6, 7). The use of either a fixed r2 value of 0.2 or a value estimated using the 95th percentile method resulted in similar values for LD decay over chromosomes in the three sub-populations.

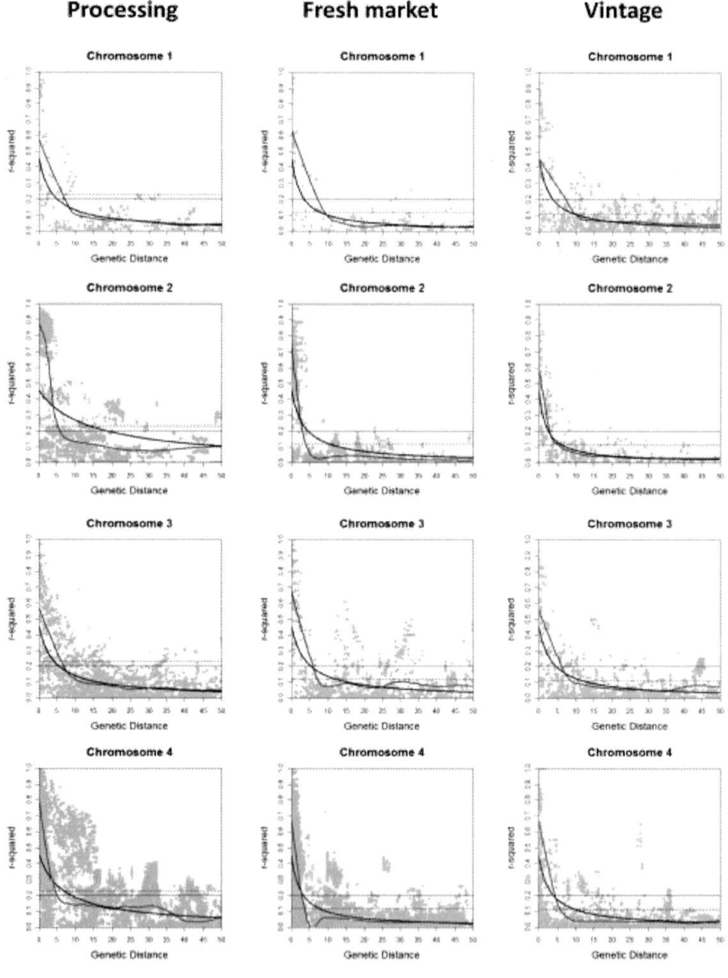

Figure 5. Linkage disequilibrium (LD) decay on chromosomes 1 to 4.

LD measures r2 against genetic map distance between pairs of SNP markers within each chromosome for processing, fresh market, and vintage subpopulations. Decay curves of locally weighted scatterplot smoothing (LOESS) [66] are represented by red, and decay curves of non-linear regression (NLR) [67] are represented by blue. Horizontal dashed and solid lines indicate the baseline r2values estimated using the 95th percentile method (0.23 for processing; 0.12 for fresh market; and 0.11 for vintage) and a fixed r2value of 0.2, respectively.

doi:10.1371/journal.pone.0045520.g005

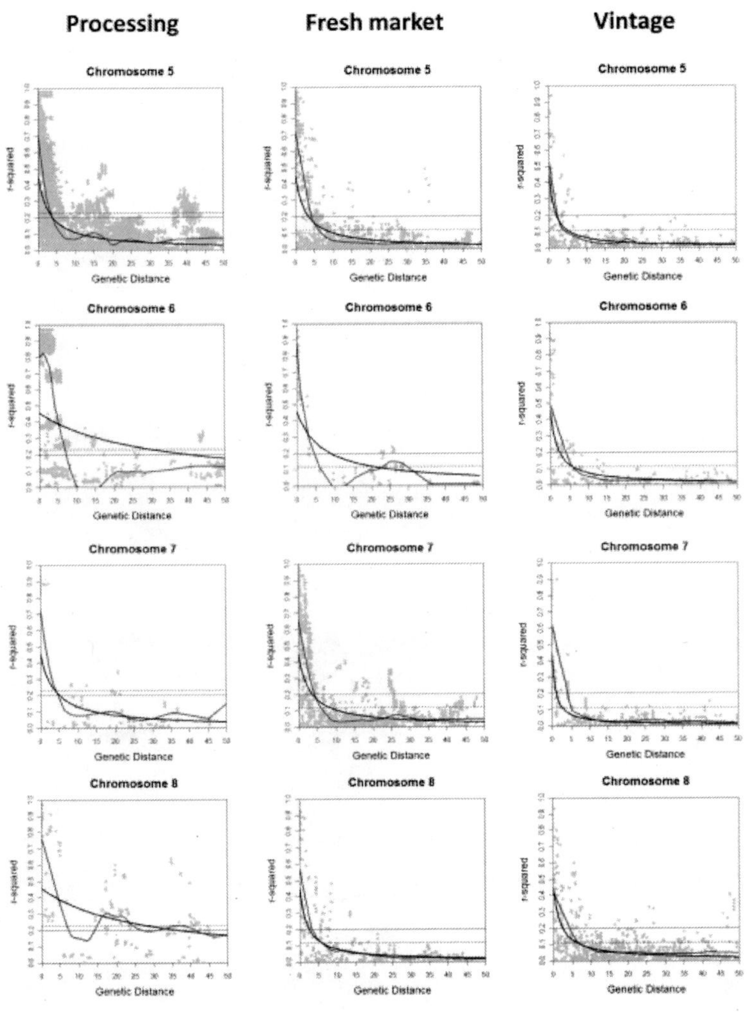

Figure 6. Linkage disequilibrium (LD) decay on chromosomes 5 to 8.

LD measures r2 against genetic map distance between pairs of SNP markers within each chromosome for processing, fresh market, and vintage subpopulations. Decay curves of locally weighted scatterplot smoothing (LOESS) [66] are represented by red, and decay curves of non-linear regression (NLR) [67] are represented by blue. Horizontal dashed and solid lines indicate the baseline r2values estimated using the 95th percentile method (0.23 for processing; 0.12 for fresh market; and 0.11 for vintage) and a fixed r2value of 0.2, respectively.

doi:10.1371/journal.pone.0045520.g006

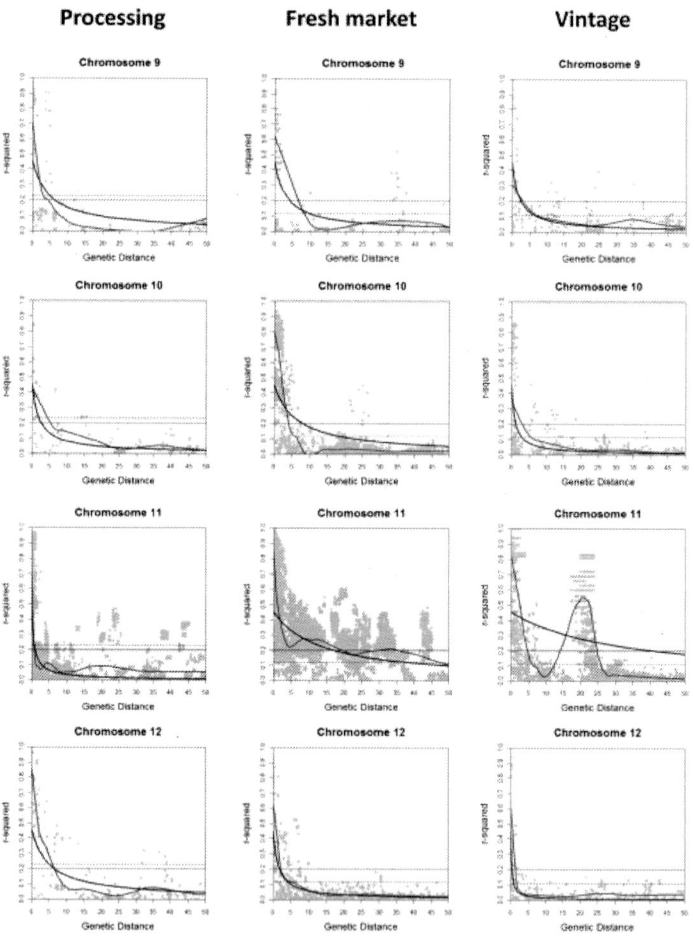

Figure 7. Linkage disequilibrium (LD) decay on chromosomes 9 to 12.

LD measures r2 against genetic map distance between pairs of SNP markers within each chromosome for processing, fresh market, and vintage sub-populations. Decay curves of locally weighted scatterplot smoothing (LOESS) [66] are represented by red, and decay curves of non-linear regression (NLR) [67] are represented by blue. Horizontal dashed and solid lines indicate the baseline r2values estimated using the 95th percentile method (0.23 for processing; 0.12 for fresh market; and 0.11 for vintage) and a fixed r2value of 0.2, respectively.

doi:10.1371/journal.pone.0045520.g007

	LD decay (cM)											
	95th percentile method[1]						Fixed method (r^2 = 0.2)					
	Processing		Fresh market		Vintage		Processing		Fresh market		Vintage	
Chr	LOESS[2]	NLR[3]	LOESS	NLR	LOESS	NLR	LOESS	NLR	LOESS	NLR	LOESS	NLR
1	6.6	4.1	9.3	7.7	9.9	9.4	7.2	5.3	7.7	3.6	7.2	3.8
2	5.4	14.2	4.1	10.2	5.8	6.6	6.0	18.7	3.3	4.6	3.0	2.7
3	6.3	4.4	7.2	12.1	8.6	11.8	6.9	5.7	6.1	5.4	6.1	4.7
4	4.6	6.9	3.8	7.4	5.8	9.9	5.2	9.0	2.9	3.5	4.4	4.1
5	3.5	3.1	6.1	7.7	3.8	4.7	3.9	4.1	4.3	3.5	2.5	1.9
6	7.1	35.2	6.0	20.1	5.8	6.0	7.4	40.6	5.0	9.7	4.1	2.8
7	4.0	4.0	6.0	8.3	4.9	2.8	5.0	6.4	4.4	3.8	4.1	1.1
8	7.7	27.9	5.7	5.7	6.7	6.3	7.7	30.7	3.3	2.8	4.4	2.5
9	3.5	5.0	9.0	9.5	7.8	6.5	4.7	6.6	6.9	4.7	4.1	2.7
10	5.2	2.2	5.8	17.7	5.8	3.6	6.0	2.8	4.2	8.0	3.6	1.4
11	0.8	0.8	47.6	38.0	6.0	n.d.	0.8	1.1	19.7	17.2	4.9	41.3
12	6.7	5.5	4.6	4.1	1.9	1.1	6.7	6.7	2.2	1.9	1.6	0.5
Average	5.0	9.8	10.1	13.3	6.4	6.8	5.5	12.5	6.2	6.1	4.4	6.3

[1]The estimate of the 95th percentile baseline r2 value in each germplasm group was 0.23 in the processing varieties, 0.12 in the fresh market varieties, and 0.11 in the vintage varieties.

[2]Logally weighted scatterplot smoothing [66].

[3]Non-linear regression [67]. For NLR, the expected r2 was calculated using the model of Hill and Weir (1988).

n.d. = not determined.

doi:10.1371/journal.pone.0045520.t005

Table 5. Chromosome by chromosome linkage disequilibrium (LD) analysis within three representative sub-populations of cultivated tomato.

doi:10.1371/journal.pone.0045520.t005

Different patterns of LD decay were observed between chromosomes and sub-populations (Table 5 and Figures 5, 6, 7). Baseline r2 values estimated using the 95th percentile method ranged from 0.11 to 0.23. In the processing sub-population, the baseline r2 value based on the 95th percentile method was 0.23 and led to estimates of LD decay that ranged from 0.8 cM on chromosome 11 (both LOESS and NLR) to 7.7 cM on chromosome 8 (LOESS) and 35.2 cM on chromosome 6 (NLR) (Table 5 and Figures 5, 6, 7). LD decay estimated from the 95th percentile baseline r2 value of 0.12 in the fresh market sub-population ranged from 3.8 cM on chromosome 4 to 47.6 cM on chromosome 11 for LOESS, and ranged from 4.1 cM on chromosome 12 to 38 cM on chromosome 11 using NLR. With the 95th percentile baseline r2 value of 0.11 in the vintage sub-population, LD decay occurred over the shortest distance on chromosome 12 (1.9 cM for LOESS and 1.1 cM for NLR), while the greatest distance for LD decay was found on chromosome 1 (9.9 cM for LOESS) and chromosome 3 (11.8 cM for NLR) (Table 5 and Figures 5, 6, 7). Estimates of LD decay vary more on a chromosome to chromosome basis than those based on the method used to establish the LD cut-off or decay curve.

DISCUSSION

A high density array with 7,720 SNP markers was used to genotype the SolCAP germplasm panel consisting of 410 inbred accessions. This diverse germplasm enabled the development of a cluster file for accurate SNP calling with the GenomeStudio software (Illumina Inc. San Diego, CA, USA). Over 98% of SNP markers generated consistent calls between duplicate samples across 34 accessions, with differences due mostly to DNA quality. Also, 7,375 SNP markers (96%) were polymorphic with <10% missing data in the entire germplasm. We found that the level of heterozygosity (proportion of heterozygotes) was low, but variable between market classes of germplasm.

The SNPs provide excellent genome coverage, but their distribution is not reflective of chromosome size. For example, chromosome 1 is cytologically one of the largest, yet is underrepresented by SNPs relative to chromosome 11 which is cytologically small. It is possible that this distribution represents a distortion of the true measure of polymorphism, heterozygosity, or genetic diversity due to the sampling of markers. However, the germplasm presented in this study were well represented in the sequencing and SNP discovery pipeline [10] with five of the seven sub-populations contributing to sequenced germplasm. We found more variation in the level of polymorphism between chromosomes within a sub-population relative to the variation between sub-populations. These results suggest that there is little ascertainment bias within the red-fruited species, and that the chromosome to chromosome variation reflects breeding history rather than polymorphism discovery. The possibility of ascertainment bias when applied to more distant germplasm exists, and will be the subject of a future study.

Cultivated germplasm was divided into distinct sub-populations. The genetic differentiation between processing and fresh market germplasm reflects human selection for distinct ideotypes tailored to the needs of specific production systems. The contemporary processing and fresh-market germplasm were also distinct from vintage germplasm, which is consistent with previous findings [6]-[9]. The further division within processing germplasm confirms an earlier analysis based on the Bayesian model implemented in STRUCTURE [68] which demonstrated sub-structure consistent with breeding history and environmental adaptation [9]. We have not seen evidence for sub-structure within fresh market germplasm previously [9]. Upon close inspection, the cluster demarked by PC1 coordinates (-10, 10) and PC2 coordinates (−30, −15) contained 87% of the fresh market accessions from Florida and 65% of the accessions from North Carolina (Figure 1B). In contrast, the cluster from PC1 (10, 30) and PC2 (-15, 10) contained 100% of the fresh market accessions from Oregon and California. The Oregon and California accessions were not part of our previous collection [9], and the sub-structure identified in this analysis suggests that diversifying forces may play a role in shaping fresh market as well as processing tomato germplasm. Genetic differentiation within market classes may reflect founder effects in breeding programs, selection for specific traits, environmental adaptation, or a combination of these factors. Cultivated cherry, wild cherry and landrace accessions were not well separated. These results support previous studies showing that Latin American accessions and feral accessions often share alleles [69]. The lack of differentiation between cultivated cherry and wild cherry or landrace accessions was somewhat surprising, but is consistent with a diverse breeding base for the cherry market class.

Genetic diversity for each sub-population was measured using allelic richness, expected heterozygosity, and polymorphic information content. The descriptive statistics in the contemporary sub-populations exceeded levels found in the vintage sub-population, but were lower relative to the cherry sub-populations and S. pimpinellifolium. Although differences between processing and vintage sub-populations may be population size dependent, rarefaction analysis provides strong support for increased variation in fresh-market sub-population relative to vintage sub-population. Tomato has undergone genetic bottlenecks during domestication and through selection after the introduction of the crop into Europe [2], [70]. Since the early 1900s, wild relatives have been used to introgress new alleles into cultivated tomato [3]–[5]. This practice is expected to increase allelic diversity in contemporary processing and large-fruited fresh-market germplasm.

The analysis of minor allele frequency (MAF) across the genome demonstrates genetic differentiation between sub-populations in specific chromosome regions. The different MAF patterns between three cultivated sub-populations (processing, fresh market, and vintage) were particularly evident on chromosomes 2, 4, 5, 6, and 11. The same regions of the genome were highlighted by SNPs contributing high absolute values to PCA loadings and by SNPs detected as Fst outliers based on a deviation from the expected distribution of Fst and He. Thus, these chromosomes appear to be under diversifying selection relative to other regions of the genome.

In order to investigate potential regions of the genome under selection, we superimposed MFA patterns, PCA loadings, and SNPs identified from Fst outlier detection with the position of genes affecting fruit shape, size and color, disease resistance, and plant morphology (Figure 3 and Figure 4). Taken together, several regions of the genome containing candidate genes which may be under selection were detected based on coincident MAF patterns, PC loadings, or Fst outlier analysis. However, the resolution of this analysis does not provide unequivocal evidence that selection for these candidates explains variation between sub-populations. Given our marker resolution and observed LD decay, candidate gene analysis was not highly informative. In addition, our ability to detect outliers may be influenced by allele frequencies and distribution across the sub-populations. In general, several regions under selection are compatible with the introgression of genes for fruit size and shape (e.g. lc and FAS), fruit color (e.g. gf and u), disease resistance (e.g.Cf-2 and I-2), and plant morphology (e.g. s and SP5).

Our results also suggest several chromosomes or regions of chromosomes with no obvious candidate genes to explain the observed differentiation. Chromosome 4 was identified as highly important for genetic differentiation between the sub-populations of cultivated germplasm. The role of this chromosome is less clear, though multiple loci affecting plant habit (e.g. dmt, Epi, glo, and si) have been mapped relative to morphological markers on this chromosome (http://tgrc.ucdavis.edu). Other areas under selection that are poorly explained by candidate genes include the chromosome 5 centromere which differentiates processing germplasm from the other sub-populations.

Common alleles at many loci in the S. pimpinellifolium accessions are minor alleles in the other sub-populations, including cherry tomato. A recent study with 144 cherry accessions demonstrated a separation into two groups: one close to the cultivated tomato and one showing admixture of cultivated tomato and S. pimpinellifolium [71]. Genetic diversity was higher in the S. pimpinellifolium relative to the wild cherry and cultivated tomato accessions, a result that is consistent with our findings. We expect accessions of S. pimpinellifolium to be relatively diverse due to their range of mating systems (many accessions exhibit facultative outcrossing) and wide geographic distribution in the native region [72]. The wild cherry accessions may be a mixture of feral cultivated accessions and wild progenitors of cultivated tomato [71].

The decay of LD over genetic distance is important to determine the density of markers appropriate for genetic analysis and selection strategies. LD levels vary both within and between species [73]. Previous estimates of LD decay in tomato were based on the entire genome with an average of 6–14 cM in processing accessions [74], 3–16 cM in fresh market accessions [74], and 15–20 cM in commercial European greenhouse accessions [75]. Given the marker density across each chromosome, we estimated LD decay on a chromosome by chromosome basis for processing, fresh market, and vintage accessions. LD decay was variable between chromosomes and sub-populations, suggesting that historical recombination is not uniform across the genome. For example, chromosome 11 showed decay over 0.8 cM for the processing sub-population, and decay over 19.7 cM for the fresh market sub-population. Although the similar estimates of LD decay was found between the LOESS and NLR methods on most of chromosomes, high levels of difference were found on chromosomes 6, 8, or 11 depending on the sub-populations. These chromosomes also showed high levels of non-homogenous distributions for pairwise r2 values relative to the other chromosomes. This variation may reflect structure within sub-populations due to selection. When the a priori vintage germplasm sub-population was based on age of variety, regardless of fruit size or geographical origin, the LD decay pattern for chromosome 4 displayed extensive LD and what appeared to be parallel patterns of decay. When the vintage accessions were separated into large fruited vintage and cultivated cherry/landrace groups, the dual pattern of LD decay disappeared and LD was reduced (Figure S1).

The patterns of LD decay we observed appear to be a consequence of introgression and directional selection through breeding. The range of LD decay also suggests that recombination remains limiting in cultivated tomato because of the inbreeding mating system and an emphasis on backcrossing and pedigree selection in tomato breeding programs [22], [76], [77]. With LD decay exceeding 1cM, the SolCAP tomato array will be useful for most selection strategies and association studies as the average marker intervals range from 0.8 to 1.6 cM in the tomato genetic maps [19].

High-throughput SNP genotyping has provided a means to both visualize and quantify the effect of human selection on the tomato genome. Selection has reduced genetic diversity in vintage cultivated forms relative to wild and feral forms, and has changed the frequency of predominant alleles. In contrast to domestication, contemporary breeding has increased allelic diversity and heterozygosity in populations relative to vintage tomatoes. Selection has also led to sub-populations which are characterized by distinct haplotype blocks, patterns of allelic diversity, and recombination history. This analysis has highlighted specific regions of the genome that appear to be under selection, with SNPs on chromosomes 2, 4, 5, 6, and 11 clearly distinguishing fresh market and processing lineages. Our findings are not surprising given the history of breeding activities. However, incorporating this information into future breeding strategies offers both challenges and potential for creativity. Going forward, we will want to balance a desire to promote recombination and generate new combinations of alleles, with the need to preserve desirable combinations of genes. These analyses also highlight a role for several chromosomes in the differentiation of cultivated tomatoes. A role for chromosome 4 has not been highlighted by previous studies yet this chromosome appears to have regions that have been selected during breeding activities.

MATERIALS AND METHODS

Plant Material

A collection of 426 tomato accessions, referred to as the Solanaceae Coordinated Agricultural Project (SolCAP) germplasm panel, was assembled from the National Plant Germplasm System (NPGS), the C. M. Rick Tomato Genetics Resource Center (TGRC), and from public plant breeding programs. Accessions from eight universities in the United States and Canada were represented including Cornell University (USA), North Carolina State University (USA), Ohio State University (USA), Oregon State University (USA), Pennsylvania State University (USA), University of Florida (USA), University of California-Davis (USA), and Ridgetown College, University of Guelph (Canada). The germplasm panel represented only the red-fruited species S. lycopersicum and S. pimpinellifolium, and included 141 processing, 110 fresh market, 61 vintage, 27 cultivated cherry, 12 landrace, 43 wild cherry, 16 S. pimpinellifolium, and 16 hybrid accessions (Table S1). Processing and fresh market germplasm represented contemporary accessions, while vintage accessions (sometime referred to as heirlooms) represented early tomato selections that in some cases predate the application of Mendelian principles to crop improvement [7], [9]. Cherry tomatoes, often referred to as S. lycopersicum 'cerasiforme', in our germplasm panel were separated into two sub-populations: cultivated and wild accessions. Landraces included Latin American cultivars and represent early domesticates from regions near the centers of origin and domestication. The collection also contained germplasm that was considered commercially relevant, with several inbred lines that are parents of commercial hybrids [78]–[80]. The collection also included the parents of several important recombinant inbred and inbred backcross populations [45], [81]–[84], segmental substitution lines [85], and a mutation library [86]. Finally, two accessions that have been the subject of public sequencing efforts, Heinz 1706 and LA1589, were also included.

SNP Genotyping

The SolCAP germplasm panel was genotyped using a tomato array with 7,720 SNPs as implemented in the Infinium assay (Illumina Inc., San Diego, CA, USA). Details of the SolCAP SNP discovery pipeline are described previously [10], as are details of the array [19]. In addition, all SNPs on the array have been incorporated into the SGN database [87], the SNP annotation file is available through the SolCAP website [88], and sequences are available through the National Center for Biotechnology Information Sequence Read Archive (accession number SRP007969).

For each accession of the SolCAP germplasm, genomic DNA was isolated from fresh young leaf tissue according to a modified CTAB method [82]. Double-stranded DNA concentrations were quantified using the PicoGreen assay (Life Technologies Corp., Grand Island, NY, USA) and normalized to 50 ng/ul with 10 mM Tris-HCl pH 8.0, 1 mM EDTA. Genotyping was conducted with 250 ng of DNA per accession following the manufacturer's protocol for the Infinium assay. For SNP calls, the resulting intensity data were loaded in GenomeStudio version 1.7.4 (Illumina Inc., San Diego, CA, USA). In order to determine SNP genotype, we first used the automated cluster algorithm to generate initial calls. Clustering for every SNP was assessed by visual inspection and modified when the default clustering was not clearly defined. Particular attention was paid to a clear definition of the boundaries for heterozygote calls, which were reduced manually for a number of SNPs in order to reduce the number of ambiguous calls. As a result, the rate of alleles with no call was increased slightly.

Data Analysis

Physical positions of 7,666 SNPs were previously determined relative to the tomato genome sequence [19]. The SNPs with physical positions were filtered based on polymorphism and missing data. The 7,323 polymorphic SNPs with <10% missing data included 13 markers with inconsistent chromosome assignments between physical and genetic map positions. We removed these SNPs and used the 7,310 SNPs for minor allele frequency (MAF) and rarefaction analyses. For principal component analysis (PCA), pairwise Fst, descriptive statistics, and Fst outlier detection, we used 4,393 polymorphic SNPs (excluding the 13 markers) with <10% missing data that were genetically mapped in the Moneymaker x LA0121 F2 population of 184 plants [19]. A second set of 3,473 markers was selected based on genetic position in the EXPEN 2000 genetic map [19] and used for pairwise estimation of Fst. Polymorphism of these markers was determined based on the observation of at least one alternative allele in the 410 inbred accessions. For linkage disequilibrium (LD) analysis, the 4,393 SNP markers were filtered for a MAF of ≥10% within each sub-population.

Principal component analysis

The genetic relationship between sub-populations was analyzed using PCA as implemented in R [89]. GenomeStudio SNP data were converted to proportional scoring where 2 is equal to homozygous for the common allele; 1 is equal to heterozygote; and 0 is homozygous for the rare allele. Missing data were imputed using the R package pcaMethods [90] in which missing data calls are based on the SVDimpute algorithm [91]. PCA was conducted for the entire data set, and subsequently for a data set consisting of only the three major sub-populations of cultivated tomato (processing, fresh market, and vintage). The relationship between accessions was visualized by plotting scores for PCs. Marker contributions to the loadings of each PC were extracted and displayed relative to chromosome position in order to visualize regions of the genome containing markers that maximize variation within the germplasm collection and assuming that these represent regions of the genome with maximum diversity. Significant differences between sub-populations were tested via analysis of variance (ANOVA) for the eigenvalues of the first two principal components.

Genetic differentiation and diversity

As a measure of population differentiation, pairwise Fst [92] was estimated using the Microsatellite analyzer v4.05 [93]. This analysis was conducted using two sets of markers (3,473 and 4,393 SNPs) for all 410 inbred accessions. We also estimated pairwise Fst with a reduction of sample size (n = 40) for processing, fresh market, vintage sub-populations in order to estimate Fst without bias due to different population sizes. The analysis was iterated for three separate sets of accessions with n = 40 for each market class. The P-value for the pairwise Fst was obtained from 10,000 permutations of genotypes and a Bonferroni correction was applied. Genetic diversity within each sub-population was assessed based on allelic richness (A) [94], [95], expected heterozygosity (He) [96] and polymorphic information content (PIC) [97] using the 4,393 SNPs for the 410 accessions. A and He were estimated using the MSA software which corrects for sample size [98], while PIC was calculated using the equation:

$$PIC = 1 - \sum_{i=1}^{n} p_i^2 - \sum_{i=1}^{n-1} \sum_{j=i+1}^{n} 2p_i^2 p_j^2$$

where n is the number of allele and pi is the frequency of the ith allele [97].

Rarefaction analysis

This analysis was used to estimate how the number of polymorphic markers increases relative to sample size [26]. Analysis and graphing was conducted using the species accumulation curve function "specaccum" in the R package, vegan [99]. The method "random" was applied to estimate means and standard deviations from 100 sub-samples without replacement [26]. This method provides a data summary and permits boxplot methods to be used to graph the curves. The boxplot function was used to superimpose standard deviations.

Minor allele frequency

The minor allele was determined based on the allele calls for 410 inbred accessions. Minor allele frequency (MAF) was then calculated across all chromosomes for each of sub-populations and sub-groups within the processing and fresh market sub-populations. This approach permitted the comparison of allele frequencies between sub-populations. MAF was graphed to visualize genetic variation based on physical map position across 12 chromosomes using the R package ggplot2 [100]. We also positioned genes for disease resistance (13 genes), fruit shape and size (5 genes), fruit color (5 genes), and plant morphology (5 genes) on the MAF plot (Table S9). Physical positions of the genes relative to the tomato genome sequences were determined using coding sequences or flanking marker sequences. These sequences were aligned to the Tomato WGS chromosome v SL2.40 [101], [102] using BLAST [103].

Loci under positive selection

In order to identify loci under positive selection between processing, fresh market, and vintage sub-populations, we used an Fst- outlier detection method as implemented in the LOSITAN workbench [27], [28]. Outliers were detected based on a distribution of Fst and expected heterozygosity (He). Five simulations for each pairwise comparison were run with 10,000 iterations. Simulations were conducted using the neutral mean, forced mean, and a mutation model with infinite alleles. Under these options, LOSITAN estimated a 95% confidence interval, and SNPs that fell outside of this range were identified as under positive selection (higher than expected Fst for an estimated He).

Linkage disequilibrium (LD) analysis

The extent of LD across each chromosome was measured using the SNP genotypes in three representative groups of cultivated tomatoes (processing, fresh market, and vintage). Pairwise r2 between markers within each chromosome was then calculated using TASSEL v 2.1 [104] and GGT v 2 [105]. In order to determine the decay of LD, these r2 values were plotted against the genetic distance (cM) between markers. Curves of LD decay were fitted using locally weighted scatterplot smoothing (LOESS) [66] and non-linear regression (NLR) [67] using R [89]. For LOESS, smoothing parameters between 0.1 and 0.5 were tested, and a final parameter of 0.3 was chosen based on curve fits. For NLR, the predicted r2 values between adjacent markers were calculated using the model of Hill and Weir [106]. Two methods were chosen to determine baseline r2 values: a fixed value of 0.2 [107] and the parametric 95th percentile of the distribution of the unlinked markers [108].

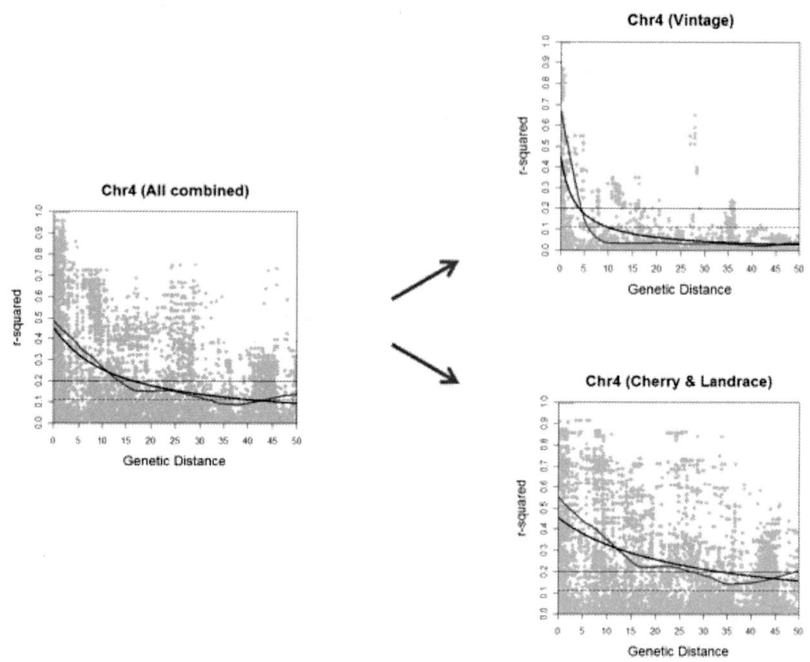

Figure S1.

Linkage disequilibrium (LD) decay on chromosome 4 for vintage, cultivated cherry, and landrace accessions. LD measures r2 against genetic map distance between pairs of SNP markers. Decay curves are represented by red (LOESS) and blue (non-linear regression). The baseline r2values were indicated by horizontal dashed line (the 95th percentile r value of 0.11) and solid line (the fixed r2 value of 0.2).

doi:10.1371/journal.pone.0045520.s001

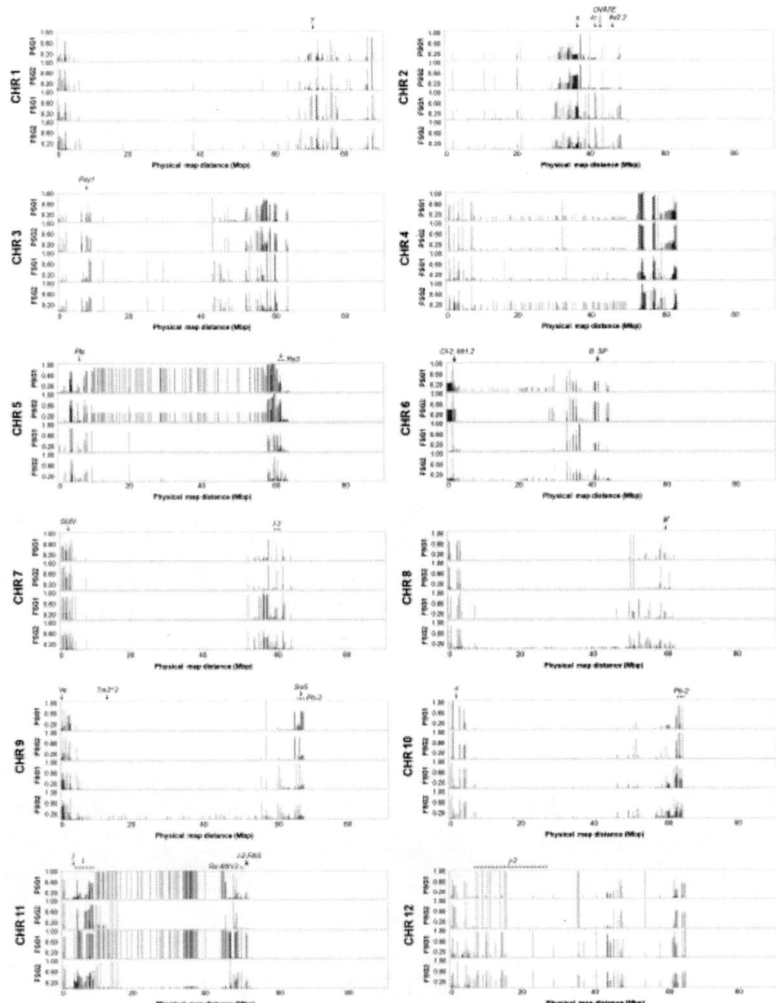

Figure S2.

Minor allele frequency (MAF) for further divisions within both processing and fresh market germplasm. The minor allele was determined relative to allele calls for all 410 inbred accessions based on 7,310 SNPs, and then MAF was estimated and graphed within each sub-group. Proc 1 and Proc 2 indicate sub-groups of processing germplasm, and FM 1 and FM2 indicate sub-groups of fresh market germplasm.

doi:10.1371/journal.pone.0045520.s002

ACKNOWLEDGMENTS

We would like to thank Esther Van der Knaap at The Ohio State University, OARDC; the C. M. Rick Tomato Genetics Resource Center (TGRC); and the National Plant Germplasm System (NPGS) for providing tomato seeds. We thank Carolyn Male for discussion and suggestions of representative heirloom accessions and Joanne Labate for helpful discussion. We also thank internal reviewers at The Ohio State University, OARDC for comments and helpful suggestions on the manuscript.

Author Contributions

Conceived and designed the experiments: SCS AVD RTC SFH JWS RGG DRP MM JRM DMF. Performed the experiments: SCS AVD KS DSD DZ DMF. Analyzed the data: SCS MWG DMF. Contributed reagents/materials/analysis tools: MWG RTC SFH JWS RGG DRP MM JRM DMF. Wrote the paper: SCS MWG DMF.

REFERENCES

1. Tanksley SD, McCouch SR (1997) Seed banks and molecular maps: Unlocking genetic potential from the wild. Science 277: 1063–1066. doi: 10.1126/science.277.5329.1063
2. Miller JC, Tanksley SD (1990) RFLP analysis of phylogenetic relationships and genetic variation in the genus *Lycopersicon*. Theor Appl Genet 80: 437–448. doi: 10.1007/bf00226743
3. United States Department of Agriculture, States Relations Service, Office of Experiment Stations (1919) Experiment Station Record Vol 39. UNT Digital Library. Available:http://digital.library.unt.edu/ark:/67531/metadc5015/. Accessed 2012 Aug 27.
4. Alexander LJ (1934) Leaf mold resistance in the tomato. Ohio Agri Exp Sta Bull 539.
5. Watts VM (1947) The use of *Lycopersicon peruvianum* as a source of nematode resistance in tomatoes. Proc Amer Soc Hort Sci 49: 233–234.
6. Park YH, West MAL, St Clair DA (2004) Evaluation of AFLPs for germplasm fingerprinting and assessment of genetic diversity in cultivars of tomato (*Lycopersicon esculentum* L.). Genome 47: 510–518. doi: 10.1139/g04-004

7. Williams CE, St. Clair DA (1993) Phenetic relationships and levels of variability detected by restriction fragment length polymorphism and random amplified polymorphic DNA analysis of cultivated and wild accessions of *Lycopersicon esculentum*. Genome 36: 619–630. doi: 10.1139/g93-083
8. Sim SC, Robbins MD, Chilcott C, Zhu T, Francis DM (2009) Oligonucleotide array discovery of polymorphisms in cultivated tomato (*Solanum lycopersicum* L.) reveals patterns of SNP variation associated with breeding. BMC Genomics 10: 10. doi: 10.1186/1471-2164-10-10
9. Sim SC, Robbins MD, Van Deynze A, Michel AP, Francis DM (2011) Population structure and genetic differentiation associated with breeding history and selection in tomato (*Solanum lycopersicum* L.). Heredity 106: 927–935. doi: 10.1038/hdy.2010.139
10. Hamilton JP, Sim S, Stoffel K, Van Deynze A, Buell CR, et al. (2012) Single nucleotide polymorphism discovery in cultivated tomato via sequencing by synthesis. The Plant Genome 5: 17–29. doi: 10.3835/plantgenome2011.12.0033
11. Gupta PK, Rustgi S, Mir RR (2008) Array-based high-throughput DNA markers for crop improvement. Heredity 101: 5–18. doi: 10.1038/hdy.2008.35
12. Yang W, Bai XD, Kabelka E, Eaton C, Kamoun S, et al. (2004) Discovery of single nucleotide polymorphisms in *Lycopersicon esculentum* by computer aided analysis of expressed sequence tags. Mol Breeding 14: 21–34. doi: 10.3835/plantgenome2011.12.0033
13. Labate JA, Baldo AM (2005) Tomato SNP discovery by EST mining and resequencing. Mol Breeding 16: 343–349. doi: 10.3835/plantgenome2011.12.0033
14. Jiménez-Gómez J, Maloof J (2009) Sequence diversity in three tomato species: SNPs, markers, and molecular evolution. BMC Plant Biol 9: 85. doi: 10.1186/1471-2229-9-85
15. Van Deynze A, Stoffel K, Buell CR, Kozik A, Liu J, et al. (2007) Diversity in conserved genes in tomato. BMC Genomics 8: 465. doi: 10.1186/1471-2164-8-465
16. Labate JA, Robertson LD, Wu FN, Tanksley SD, Baldo AM (2009) EST, COSII, and arbitrary gene markers give similar estimates of nucleotide diversity in cultivated tomato (*Solanum lycopersicum* L.). Theor Appl Genet 118: 1005–1014. doi: 10.1007/s00122-008-0957-2
17. Steemers FJ, Chang WH, Lee G, Barker DL, Shen R, et al. (2006) Whole-genome genotyping with the single-base extension assay. Nat Methods 3: 31–33. doi: 10.1038/nmeth842

18. Ganal MW, Durstewitz G, Polley A, Berard A, Buckler ES, et al. (2011) A large maize (*Zea mays* L.) SNP genotyping array: development and germplasm genotyping, and genetic mapping to compare with the B73 reference genome. PLoS ONE 6: e28334. doi: 10.1371/journal.pone.0028334

19. Sim S, Durstewitz G, Plieske J, Wieseke R, Ganal M, et al. (2012) Development of a large SNP genotyping array and generation of high-density genetic maps in tomato. PLoS ONE 7: e40563. doi: 10.1371/journal.pone.0040563

20. Zhao K, Tung CW, Eizenga GC, Wright MH, Ali ML, et al. (2011) Genome-wide association mapping reveals a rich genetic architecture of complex traits in *Oryza sativa*. Nat Commun 2: 467. doi: 10.1038/ncomms1467

21. Thomson MJ, Zhao K, Wright M, McNally KL, Rey J, et al. (2012) High-throughput single nucleotide polymorphism genotyping for breeding applications in rice using the BeadXpress platform. Mol Breeding 29: 875–886. doi: 10.1007/s11032-011-9663-x

22. Graham TO (1959) Impact of recorded mendelian factors on the tomato 1929–1959. Tom Gen Coop Rep 9: 37. doi: 10.1007/s11032-011-9663-x

23. Rasmussen WD (1968) Advances in American agriculture: The mechanical tomato harvester as a case study. Technol Cult 9: 531–543. doi: 10.1007/s11032-011-9663-x

24. Sim SC, Merk HL, McQueen J (2011) Downstream analysis of SNPs from the SolCAP tomato infinium aray (II). Available: http://www.extension.org/pages/61007. Accessed 2012 Aug 27.

25. Labate JA, Francis D, McGrath MT, Panthee D, Robertson LD (2010) Diversity in a collection of heirloom tomato varieties. HortScience 45: S145–S145.

26. Gotelli NJ, Colwell RK (2001) Quantifying biodiversity: procedures and pitfalls in the measurement and comparison of species richness. Ecol Lett 4: 379–391.

27. Antao T, Lopes A, Lopes RJ, Beja-Pereira A, Luikart G (2008) LOSITAN: A workbench to detect molecular adaptation based on a Fst-outlier method. BMC Bioinformatics 9: 323. doi: 10.1186/1471-2105-9-323

28. Beaumont MA, Nichols RA (1996) Evaluating loci for use in the genetic analysis of population structure. Proc R Soc B 263: 1619–1626.

29. Frary A, Nesbitt TC, Frary A, Grandillo S, van der Knaap E, et al. (2000) *fw2.2*: a quantitative trait locus key to the evolution of tomato fruit size. Science 289: 85–88. doi: 10.1126/science.289.5476.85

30. Van der Hoeven R, Ronning C, Giovannoni J, Martin G, Tanksley S

(2002) Deductions about the number, organization, and evolution of genes in the tomato genome based on analysis of a large expressed sequence tag collection and selective genomic sequencing. Plant Cell 14: 1441–1456. doi: 10.1105/tpc.010478

31. Liu JP, Van Eck J, Cong B, Tanksley SD (2002) A new class of regulatory genes underlying the cause of pear-shaped tomato fruit. Proc Natl Acad Sci U S A 99: 13302–13306. doi: 10.1073/pnas.162485999

32. Munos S, Ranc N, Botton E, Berard A, Rolland S, et al. (2011) Increase in tomato locule number is controlled by two single-nucleotide polymorphisms located near WUSCHEL. Plant Physiol 156: 2244–2254. doi: 10.1104/pp.111.173997

33. Xiao H, Jiang N, Schaffner E, Stockinger EJ, van der Knaap E (2008) A retrotransposon-mediated gene duplication underlies morphological variation of tomato fruit. Science 319: 1527–1530. doi: 10.1126/science.1153040

34. Cong B, Barrero LS, Tanksley SD (2008) Regulatory change in YABBY-like transcription factor led to evolution of extreme fruit size during tomato domestication. Nat Genet 40: 800–804. doi: 10.1038/ng.144

35. Solanaceae Coordinated Agricultural Project website. Available:http://solcap.msu.edu/tomato_phenotype_data.shtml. Accessed 2012 Aug 27.

36. Sol Genomics Network website. Availalbe: http://solgenomics.net/search/phenotypes/stock. Accessed 2012 Aug 27.

37. Merk HL, Yarnes SC, Van Deynze A, Tong N, Menda N, et al.. (2012) Trait diversity and potential for selection indices based on variation among regionally adapted processing tomato germplasm. J Amer Soc Hort Sci. In press.

38. Luo J, Butelli E, Hill L, Parr A, Niggeweg R, et al. (2008) AtMYB12 regulates caffeoyl quinic acid and flavonol synthesis in tomato: expression in fruit results in very high levels of both types of polyphenol. Plant J 56: 316–326. doi: 10.1111/j.1365-313X.2008.03597.x

39. Ray J, Moureau P, Bird C, Bird A, Grierson D, et al. (1992) Cloning and characterization of a gene involved in phytoene synthesis from tomato. Plant Mol Biol 19: 401–404. doi: 10.1007/BF00023387

40. Yuan DJ, Chen J, Shen HL, Yang WC (2008) Genetics of flesh color and nucleotide sequence analysis of phytoene synthase gene 1 in a yellow-fruited tomato accession PI114490. Sci Hortic-Amsterdam 118: 20–24. doi: 10.1016/j.scienta.2008.05.011

41. Ronen G, Carmel-Goren L, Zamir D, Hirschberg J (2000) An alternative pathway to beta-carotene formation in plant chromoplasts discovered by map-based cloning of Beta and old-gold color mutations in tomato.

Proc Natl Acad Sci U S A 97: 11102–11107. doi: 10.1073/pnas.190177497

42. Barry CS, McQuinn RP, Chung MY, Besuden A, Giovannoni JJ (2008) Amino acid substitutions in homologs of the STAY-GREEN protein are responsible for the green-flesh and chlorophyll retainer mutations of tomato and pepper. Plant Physiol 147: 179–187. doi: 10.1104/pp.108.118430

43. Powell A, Nguyen C, Hill T, Cheng K, Figueroa-Balderas R, et al. (2012) Uniform ripening encodes a Golden 2-like transcription factor regulating tomato fruit chloroplast development. Science 336: 1711–1715. doi: 10.1126/science.1222218

44. Martin GB, Brommonschenkel SH, Chunwongse J, Frary A, Ganal MW, et al. (1993) Map-based cloning of a protein kinase gene conferring disease resistance in tomato. Science 262: 1432–1436. doi: 10.1126/science.7902614

45. Yang W, Sacks EJ, Ivey MLL, Miller SA, Francis DM (2005) Resistance in *Lycopersicon esculentum* intraspeciflc crosses to race T1 strains of *Xanthomonas campestris* pv.*vesicatoria* causing bacterial spot of tomato. Phytopathology 95: 519–527. doi: 10.1094/PHYTO-95-0519

46. Dixon MS, Jones DA, Keddie JS, Thomas CM, Harrison K, et al. (1996) The tomato *Cf-2*disease resistance locus comprises two functional genes encoding leucine-rich repeat proteins. Cell 84: 451–459. doi: 10.1016/S0092-8674(00)81290-8

47. Milligan SB, Bodeau J, Yaghoobi J, Kaloshian I, Zabel P, et al. (1998) The root knot nematode resistance gene *Mi* from tomato is a member of the leucine zipper, nucleotide binding, leucine-rich repeat family of plant genes. Plant Cell 10: 1307–1319. doi: 10.2307/3870642

48. Scott JW, Agrama HA, Jones JP (2004) RFLP-based analysis of recombination among resistance genes to Fusarium wilt races 1, 2, and 3 in tomato. J Amer Soc Hort Sci 129: 394–400.

49. Bohn GW, Tucker CM (1939) Immunity to fusarium wilt in the tomato. Science 89: 603–604. doi: 10.1126/science.89.2322.603

50. Pei CC, Wang H, Zhang JY, Wang YY, Francis DM, et al. (2012) Fine mapping and analysis of a candidate gene in tomato accession PI128216 conferring hypersensitive resistance to bacterial spot race T3. Theor Appl Genet 124: 533–542. doi: 10.1007/s00122-011-1726-1

51. Simons G, Groenendijk J, Wijbrandi J, Reijans M, Groenen J, et al. (1998) Dissection of the Fusarium *I2* gene cluster in tomato reveals six homologs and one active gene copy. Plant Cell 10: 1055–1068. doi: 10.2307/3870690

52. Hemming MN, Basuki S, McGrath DJ, Carroll BJ, Jones DA (2004)

Fine mapping of the tomato *I-3* gene for fusarium wilt resistance and elimination of a co-segregating resistance gene analogue as a candidate for *I-3*. Theor Appl Genet 109: 409–418. doi: 10.1007/s00122-004-1646-4

53. Lanfermeijer FC, Dijkhuis J, Sturre MJG, de Haan P, Hille J (2003) Cloning and characterization of the durable tomato mosaic virus resistance gene *Tm-2(2)* from*Lycopersicon esculentum*. Plant Mol Biol 52: 1037–1049. doi: 10.1023/A:1025434519282
54. Chunwongse J, Chunwongse C, Black L, Hanson P (2002) Molecular mapping of the *Ph-3* gene for late blight resistance in tomato. J Hortic Sci Biotech 77: 281–286.
55. Robbins MD, Masud MAT, Panthee DR, Gardner RG, Francis DM, et al. (2010) Marker-assisted selection for coupling phase resistance to tomato spotted wilt virus and*Phytophthora infestans* (late blight) in tomato. HortScience 45: 1424–1428.
56. Kawchuk LM, Hachey J, Lynch DR, Kulcsar F, van Rooijen G, et al. (2001) Tomato *Ve*disease resistance genes encode cell surface-like receptors. Proc Natl Acad Sci U S A 98: 6511–6515. doi: 10.1073/pnas.091114198
57. Spassova MI, Prins TW, Folkertsma RT, Klein-Lankhorst RM, Hille J, et al. (2001) The tomato gene *Sw5* is a member of the coiled coil, nucleotide binding, leucine-rich repeat class of plant resistance genes and confers resistance to TSWV in tobacco. Mol Breeding 7: 151–161.
58. Moreau P, Thoquet P, Olivier J, Laterrot H, Grimsley N (1998) Genetic mapping of *Ph-2*, a single locus controlling partial resistance to *Phytophthora infestans* in tomato. Mol Plant Microbe In 11: 259–269.
59. Kim MJ, Mutschler MA (2005) Transfer to processing tomato and characterization of late blight resistance derived from *Solanum pimpinellifolium* L. L3708. J Amer Soc Hort Sci 130: 877–884.
60. Lippman ZB, Cohen O, Alvarez JP, Abu-Abied M, Pekker I, et al. (2008) The making of a compound inflorescence in tomato and related nightshades. Plos Biol 6: e288. doi: 10.1371/journal.pbio.0060288
61. Pnueli L, CarmelGoren L, Hareven D, Gutfinger T, Alvarez J, et al. (1998) The *SELF-PRUNING* gene of tomato regulates vegetative to reproductive switching of sympodial meristems and is the ortholog of *CEN* and *TFL1*. Development 125: 1979–1989.
62. Jones CM, Rick CM, Adams D, Jernstedt J, Chetelat RT (2007) Genealogy and fine mapping of obscuravenosa, a gene affecting the distribution of chloroplasts in leaf veins, and evidence of selection during breeding of tomatoes (*Lycopersicon esculentum*; Solanaceae). Am J Bot 94: 935–947. doi: 10.3732/ajb.94.6.935

63. Carmel-Goren L, Liu YS, Lifschitz E, Zamir D (2003) The *SELF-PRUNING* gene family in tomato. Plant Mol Biol 52: 1215–1222. doi: 10.1023/B:PLAN.0000004333.96451.11
64. Mao L, Begum D, Chuang HW, Budiman MA, Szymkowiak EJ, et al. (2000) *JOINTLESS*is a MADS-box gene controlling tomato flower abscission zone development. Nature 406: 910–913. doi: 10.1038/35022611
65. Budiman MA, Chang SB, Lee S, Yang TJ, Zhang HB, et al. (2004) Localization of*jointless-2* gene in the centromeric region of tomato chromosome 12 based on high resolution genetic and physical mapping. Theor Appl Genet 108: 190–196. doi: 10.1007/s00122-003-1429-3
66. Cleveland WS (1979) Robust locally weighted regression and smoothing scatterplots. J Am Stat Assoc 74: 829–836. doi: 10.1046/j.1461-0248.2002.00367.x
67. Bates DM, Watts DG (1988) Nonlinear regression analysis and its applications. Indianapolis: Wiley Publishing, Inc.
68. Pritchard JK, Stephens M, Donnelly P (2000) Inference of population structure using multilocus genotype data. Genetics 155: 945–959. doi: 10.1046/j.1461-0248.2002.00367.x
69. Rick CM (1958) The role of natural hybridization in the derivation of cultivated tomatoes of western South America. Economic Bot 12: 346–367. doi: 10.1046/j.1461- 0248.2002.00367.x
70. Rick CM (1976) Tomato, *Lycopersicom esculentum* (Solanaceae). In: Simmonds NW, editor. Evolution of crop plants. London, England: Longman Group. 268–273.
71. Ranc N, Munos S, Santoni S, Causse M (2008) A clarified position for *Solanum lycopersicum* var. *cerasiforme* in the evolutionary history of tomatoes (solanaceae). BMC Plant Biol 8: 130. doi: 10.1186/1471-2229-8-130
72. George WL, Berry SA (1983) Genetics and breeding of processing tomatoes. In: Gould WA, editor. Tomato production, processing and quality evaluation. Westport: AVI Publishing Company, Inc. 48–65.
73. Stevens MA, Rick CM (1986) Genetics and Breeding. In: Athernon JG, Rudich J, editors. The Tomato Crop A Scientific Basis for Improvement. London, England: Chapman and Hall. 35–109.
74. Gardner RG (1992) 'Mountain Spring' tomato; NC 8276 and NC 84173 tomato breeding lines. HortScience 27: 1233–1234.
75. Scott JW, Baldwin EA, Klee HJ, Brecht JK, Olson SM, et al. (2008) Fla. 8153 hybrid tomato; Fla. 8059 and Fla. 7907 breeding lines. HortScience 43: 2228–2230.

Chapter 7

POPULATION GENETIC DIVERSITY AND STRUCTURE OF A NATURALLY ISOLATED PLANT SPECIES, RHODIOLA DUMULOSA (CRASSULACEAE)

Yan Hou, Anru Lou

State Key Laboratory of Earth Surface Processes and Resource Ecology, College of Life Sciences, Beijing Normal University, Beijing, China

ABSTRACT

Aims

Rhodiola dumulosa (Crassulaceae) is a perennial diploid species found in high-montane areas. It is distributed in fragmented populations across northern, central and northwestern China. In this study, we aimed to (i) measure the genetic diversity of this species and that of its populations; (ii) describe the genetic structure of these populations across the entire distribution range in China; and (iii) evaluate the extent of gene flow among the naturally fragmented populations.

Methods

Samples from 1089 individuals within 35 populations of R. dumulosa were collected, covering as much of the entire distribution range of this species within China as possible. Population genetic diversity and structure were analyzed using AFLP molecular markers. Gene flow among populations was estimated according to the level of population differentiation.

Important Findings

The total genetic diversity of R. dumulosa was high but decreased with increasing altitude. Population-structure analysis indicated that the most closely related populations were geographically restricted and occurred in close proximity to each other. A significant isolation-by-distance pattern, caused by the naturally fragmented population distribution, was observed.

At least two distinct gene pools were found in the 35 sampled populations, one composed of populations in northern China and the other composed of populations in central and northwestern China. The calculation of Nei's gene diversity index revealed that the genetic diversity in the northern China pool (0.1972) was lower than that in the central and northwestern China pool (0.2216). The populations were significantly isolated, and gene flow was restricted throughout the entire distribution. However, gene flow among populations on the same mountain appears to be unrestricted, as indicated by the weak genetic isolation among these populations.

INTRODUCTION

Habitat fragmentation is a significant threat to the maintenance of biodiversity in many terrestrial ecosystems [1]. Many studies have investigated the effects of fragmentation on the genetic diversity and population structure of plant species [2]–[4]. In general, fragmentation is expected to reduce genetic diversity and to increase interpopulation genetic divergence by restricting gene flow among fragmented populations, increasing inbreeding and increasing random genetic drift within populations [5]. However, habitat fragmentation does not always lead to reduced genetic variation [6]–[8]. In some cases, the genetic diversity of a fragmented population can be higher than that of a continuously distributed population [9]. This is because the effects of habitat fragmentation on genetic diversity and population structure can be affected by other factors, such as population size, gene flow and the time scale of fragmentation [10].

Fragmented distributions of plant populations are caused not only by human activity but also by natural factors, such as long-term, large-scale climate oscillations, topographical changes, the isolation of suitable habitats, or other ecological changes. Studies of the genetic diversity of naturally fragmented populations may not only reveal the ecological consequences of population fragmentation over long periods of time but also provide a frame of reference for predicting the consequences of habitat fragmentation by human activities [6]. Genetic analyses of naturally fragmented populations has been conducted in several studies [6], [11]–[14].

Rhodiola dumulosa (Crassulaceae) is a perennial plant species that is found in northern, central and northwestern China. R. dumulosa is suggested to be a diploid plant [15]. It uses a mixed-mating system, i.e., it is capable of self-fertilization but most often undergoes outbreeding via pollinators [16], [17]. This species lives exclusively between rocks on mountains at elevations of 1600 m to 4100 m. Wild populations of this species are uniquely adapted to scarce and highly fragmented rocky habitats. Its habituation to these 'ecological islands' has produced a naturally fragmented distribution, which may lead to the long-term genetic isolation of its populations. Most populations in our field investigation were of moderate size (around 50 to 80 individuals each population). However, we also found a few large populations, with more than 150 individuals, and very small populations, with fewer than 10 individuals.

In this study, we used AFLP analysis to assess the genetic diversity, differentiation and structure of isolated populations of R. dumulosa from different mountains across its entire distribution in China. We then used these results to address the following questions: (1) How much genetic diversity is maintained in these naturally fragmented populations of R. dumulosa? (2) How is the genetic diversity spatially structured across the entire distribution range of this species in China? (3) Is there any gene flow among these naturally fragmented populations?

Results

Four AFLP primer combinations were used to analyze 1,089 individual genotypes, resulting in 225 markers, 221 (98.22%) of which were polymorphic.

Population genetic structure

The Mantel test (Fig. 1) revealed a strong and significant positive relationship between geographical and genetic distances (r = 0.801; P<0.01) across the whole sampled region, indicating significant isolation-by-distance. We also conducted separate Mantel tests on the Donglingshan and Heyeping populations, in which populations were collected according to altitude, to determine whether isolation-by-distance also occurred on a smaller geographical scale. These results did not indicate any significant correlation between genetic differentiation and geographical distance within these populations (Donglingshan: r = 0.246; P = 0.186; Heyeping: r = 0.0852; P = 0.38).

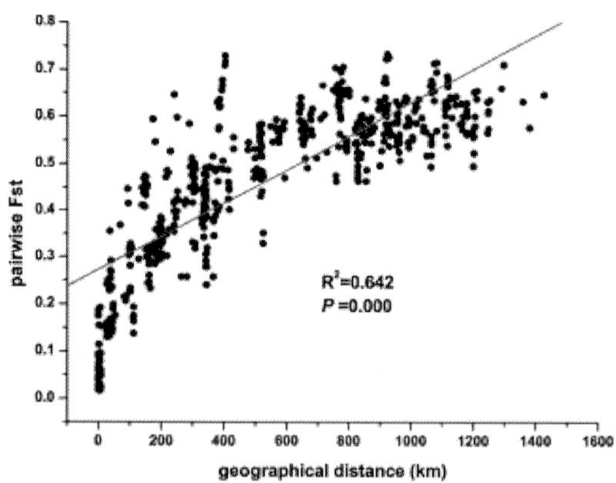

Figure 1. The correlation between pairwise Fst and pairwise geographic distance among populations of Rhodiola dumulosa.

doi:10.1371/journal.pone.0024497.g001

Hierarchical cluster analysis (UPGMA) showed that populations sampled from northern China and central China were grouped separately according to their geographical distribution (Fig. 2). Cluster I, with a bootstrap value of 60%, contained all populations in northern China, including 6 populations in Beijing, 5 populations in Hebei province, 12 populations in Shanxi province and 1 population in the Inner Mongolia autonomous region. Cluster II, with a bootstrap value of 100%, contained all populations in central China, including 1 population in Hubei province and 3 populations in Shaanxi province. The remaining northwestern populations could not be grouped in a single cluster, but the most closely related populations were geographically restricted and occurred in close proximity to each other. The NJ dendrogram revealed an obvious split between the populations in northern China and the other populations, with a bootstrap value of 99.2% (Fig. 3).

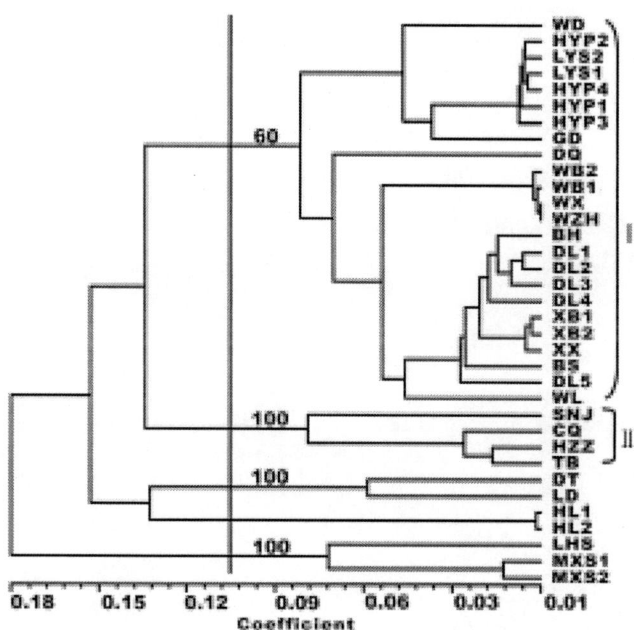

Figure 2. UPGMA tree of 35 Rhodiola dumulosa populations (numbers indicate bootstrap support values).

doi:10.1371/journal.pone.0024497.g002

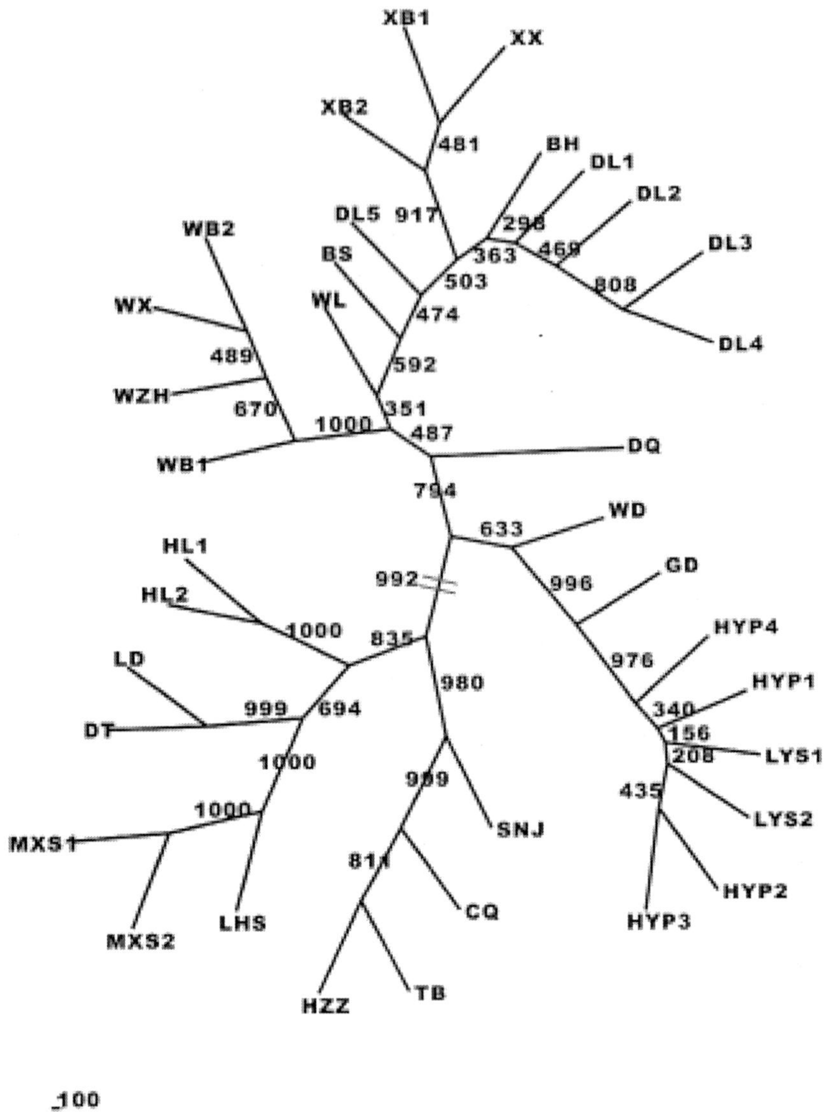

Figure 3. NJ tree of 35 Rhodiola dumulosa populations (numbers indicate bootstrap support values).

doi:10.1371/journal.pone.0024497.g003

To further test this population structure, a model-based clustering method was implemented in the program STRUCTURE [18]–[20]. Without prior information about the populations and under an admixed model, STRUCTURE calculated that the estimate of the likelihood of the data (LnP(D)) was greatest when K = 2. For K>2, LnP(D) increased slightly but more or less plateaued (Fig. 4a), i.e., ΔK reached its maximum at K = 2 (Fig. 4b), suggesting that all populations fell into one of the two clusters. These two genetically distinct clusters primarily correspond to the geographic distribution of these populations (Fig. 5), and the percent representation of each cluster in each sampled population was high (Table S1). The red cluster covered all populations in northern China, and the remaining populations were grouped in the green cluster(Fig. 4c). This result is identical to the splitting in the NJ tree. Furthermore, UPGMA Cluster I is identical to the red cluster, indicating that the grouping of northern populations is well supported. Overall, the cluster analysis strongly suggested that the 35 sampled populations can be divided into two clusters, one composed of populations in northern China (NC) and the other composed of populations in central and northwestern China (CNWC).

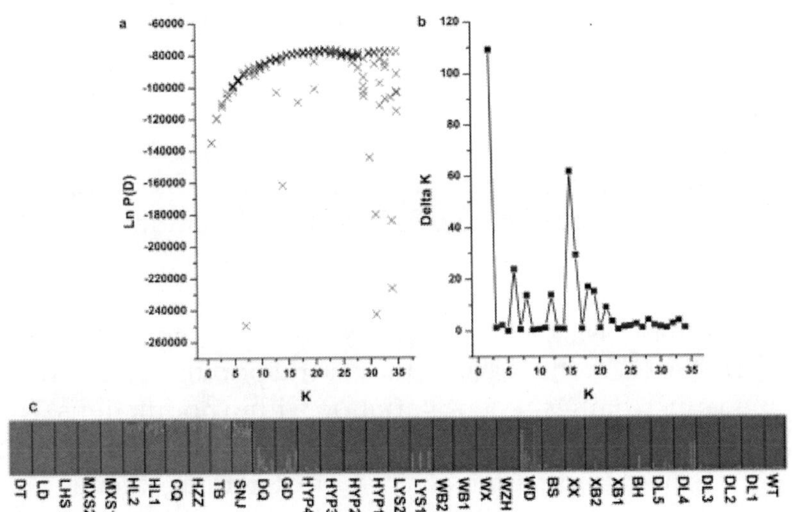

Figure 4. STRUCTURE analysis of Rhodiola dumulosa populations.

Based on AFLP data (a: the relationship between K and LnP(D); b: the relationship between K and ΔK; c: the grouping when K = 2).

doi:10.1371/journal.pone.0024497.g004

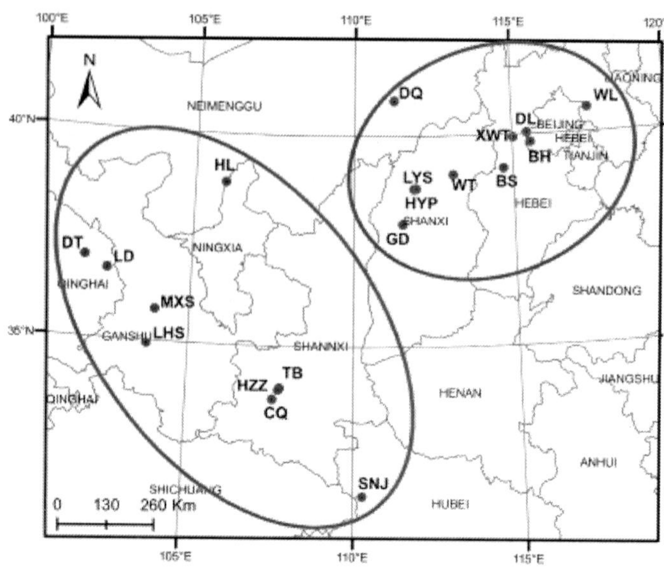

Figure 5. Distribution of the field sampling sites in China (the color of circles corresponds to two clusters resulted from STRUCTURE).

doi:10.1371/journal.pone.0024497.g005

Analysis of Molecular Variance (AMOVA) based on the two genetic clusters indicated that majority of genetic variation (43.49%) occurred within populations, while the variation between the two clusters was 23.68% (Table 1). AMOVA based on three major distribution regions gave a very similar result (42.59% variation within populations, 29.39% variation among the three regions). Moreover, AMOVA that treated the central population as one group calculated much higher genetic variation within populations (65.44%) than that among populations (34.56%), which may partly explain why the northern populations always clustered together as a group. The central populations also exhibited a higher genetic variation within populations (58.87% variation within the population and 41.13% variation among the three regions), while the northwestern populations had a slightly higher variation among populations than within populations (52.41% and 47.59%, respectively). When the central and northwestern populations were pooled together, AMOVA calculated somewhat higher genetic variation among populations (56.90%). We also treated the 35 populations as one group and compared the variation within and among populations. The Fst value was 0.49922 (P<0.001) with 50.08% within populations and 49.92% among populations, indicating that the total genetic variation was almost equally divided between intrapopulation and interpopulation variations. This approximately equal partitioning could have been the result of significant differentiation among the populations in northwestern China and frequent gene flow among some populations in northern China; we sampled more closely distributed populations in northern China than in northwestern China. Therefore, we randomly chose one population from each locality (resulting in the exclusion of populations DL2, DL3, DL4, DL5, XB1, XX, WD, WB1, WB2, WX, LYS2, HYP2, HYP3, HYP4, HL2, and MXS2) and repeated the analyses (Table 1). After excluding these populations, the Fst became 0.54252 (P<0.001) with 54.25% of variation occurring among populations, only slightly higher than the variation within populations. Therefore, we conclude that the differentiation among groups was significant, but the genetic variation within populations was maintained.

Table 1. Results of AMOVA for R. dumulosa individuals based on 225 AFLP markers.

Group	Partitioning	d.f.	Sum of squares	Variance components	Percentage of variation	F-statistics
two genetic clusters (NC & CNWC)	Among groups	1	3686.335	7.15780	23.68	Fct = 0.23679*
	Among populations within groups	33	10623.007	9.92465	32.83	Fsc = 0.43019*
	Within populations	1054	13855.386	13.14553	43.49	Fst = 0.56512*
	Total	1088	28164.728	30.22798		
three geographical regions (northern, central, northwestern)	Among groups	2	5277.625	9.07061	29.39	Fct = 0.29388*
	Among populations within groups	32	9031.718	8.64878	28.02	Fsc = 0.39684*
	Within populations	1054	13855.386	13.14553	42.59	Fst = 0.57409*
	Total	1088	28164.728	30.86492		
Northern populations	Among populations	23	5408.968	7.13192	34.56	Fst = 0.34560*
	Within populations	722	9750.255	13.50451	65.44	
	Total	745	15159.224	20.63643		
Central populations	Among populations	3	643.064	6.66237	41.13	Fst = 0.41129*
	Within populations	119	1134.839	9.53646	58.87	
	Total	122	1777.902	16.19883		
Northwestern populations	Among populations	6	2979.686	15.35977	52.41	Fst = 0.52414*
	Within populations	213	2970.292	13.94503	47.59	
	Total	219	5949.977	29.30480		
Central and Northwestern populations	Among populations	10	5214.039	16.32634	56.90	Fst = 0.56904*
	Within populations	332	4105.130	12.36485	43.10	
	Total	342	9319.169	28.69119		
19 populations from each locality	Among populations	18	8801.864	15.25624	54.25	Fst = 0.54252*
	Within populations	574	7384.328	12.86468	45.75	
	Total	592	16186.192	28.12092		

doi:10.1371/journal.pone.0024497.t001

doi:10.1371/journal.pone.0024497.t001

Population genetic diversity

The genetic diversity of each population was calculated based on the genotypes present (Table S2). The expected heterozygosity (or Nei's gene diversity, Hj) varied from 0.09690 in population MXS1 to 0.22679 in population WD. Similar numbers were calculated for the Shannon diversity index, which varied from 0.1171 in population MXS1 to 0.3174 in population WD. The total diversity of the species (Ht) was 0.2473. When the populations were divided into two clusters based on the STRUCTURE analyses, the Nei's gene diversity index was lower in the northern region of China (0.1972) than in the central and northwestern areas (0.2216). There was a significant negative correlation between population genetic diversity and altitude ($r = -0.478$, $P = 0.004$) across the entire range of R. dumulosa. A regression analysis of Nei's genetic diversity and altitude using SPSS 15.0 revealed that population diversity decreases with increasing altitude (Fig. 6).

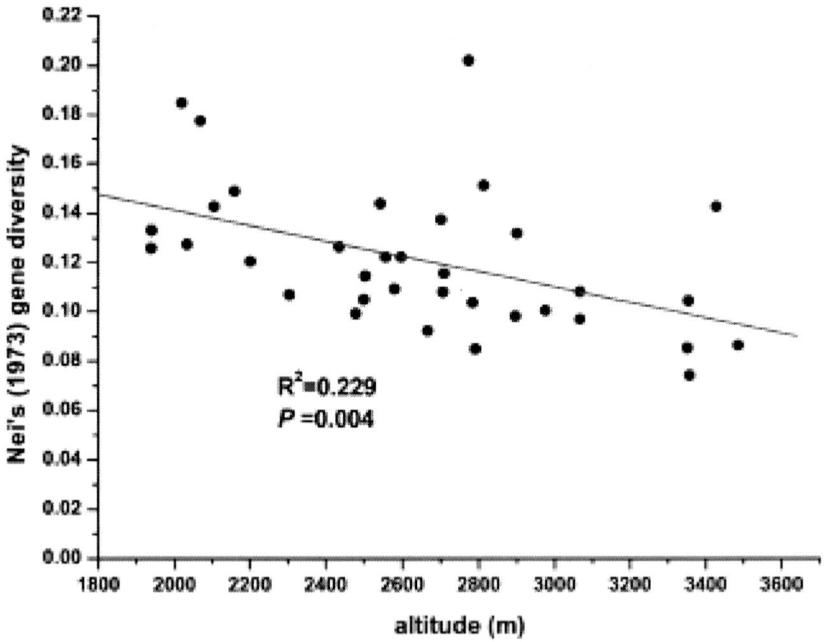

Figure 6. Scatter diagram of Nei's gene diversity and altitude.

doi:10.1371/journal.pone.0024497.g006

GENE FLOW

The genetic differentiation among 35 populations of R. dumulosa was high and significant (Fst = 0.3942, P<0.001; Gst = 0.4675). The estimated gene flow, Nm (Nm = 0.5(1− Gst)/Gst), was 0.5695. The results of this analysis reveal that the genetic differentiation among populations throughout the entire distribution area is significant and that gene flow is restricted. Furthermore, the populations in northern China always congregated together as a cluster (UPGMA tree, NJ tree and STRUCTURE analysis results). The genetic differentiation, Gst, among all northern populations was 0.3155, and the Nm was 1.0847. Thus, the gene flow among populations in northern China was much higher than that across the entire range.

On some mountains, more than one population was sampled. For instance, there were five populations in Donglingshan and four populations in Heyeping (Table 2). To investigate the gene flow among populations in these smaller geographical regions, the Gst and Nm of Donglingshan (DL1, DL2, DL3, DL4, DL5) and Heyeping (HYP1, HYP2, HYP3, HYP4) populations were calculated. The Gst of the five populations in Donglingshan was 0.1125, and the Nm was 3.9438. Meanwhile, the Gst of the four populations in Heyeping was 0.0728, and the Nm was 6.6390.

These results indicated weak genetic differentiation and frequent gene flow within these small geographical regions. Consequently, to avoid the over-estimation of gene flow caused by frequent gene flow among populations on one mountain, we calculated the Gst and Nm of 19 populations with 593 individuals from only one population from each locality (as we did for AMOVA). These results revealed that the gene flow in small regions had a clear influence (Gst19 = 0.5437, Nm19 = 0.4197) on the gene flow in the entire region (Gst = 0.4675, Nm = 0.5695).

Table 2. Localities and sizes of the 35 R. dumulosa populations sampled in this study.

Sampling Locality	Population ID	Longitude (E)	Latitude (N)	Altitude (m)	Population size
Wulingshan, Hebei province	WL	117.461°	40.577°	1939	29
Donglingshan (1), Beijing	DL1	115.454°	40.031°	2303	32
Donglingshan (2), Beijing	DL2	115.462°	40.031°	2202	32
Donglingshan (3), Beijing	DL3	115.471°	40.034°	2106	32
Donglingshan (4), Beijing	DL4	115.475°	40.033°	2018	31
Donglingshan (5), Beijing	DL5	115.451°	40.029°	2160	32
Baihuashan, Beijing	BH	115.582°	39.813°	2033	32
Xiaowutaishan (beitai 1), Hebei	XB1	115.061°	39.945°	2500	31
Xiaowutaishan (beitai 2), Hebei	XB2	115.047°	39.943°	2706	30
Xiaowutaishan (xitai), Hebei	XX	114.968°	39.912°	2480	29
Baishishan, Hebei	BS	114.698°	39.208°	1940	32
Wutaishan (dongtai), Shanxi	WD	111.662°	39.041°	2773	32
Wutaishan (zhongtai), Shanxi	WZH	113.531°	39.047°	2895	30
Wutaishan (xitai), Shanxi	WX	113.492°	39.037°	2783	29
Wutaishan (beitai 1), Shanxi	WB1	113.568°	39.080°	3066	32
Wutaishan (beitai 2), Shanxi	WB2	113.568°	39.080°	3066	28
Luyashan (1), Shanxi	LYS1	111.926°	38.750°	2701	32
Luyashan (2), Shanxi	LYS2	111.926°	38.744°	2580	30
Heyeping (1), Shanxi	HYP1	111.837°	38.722°	2709	32
Heyeping (2), Shanxi	HYP2	111.862°	38.729°	2596	32
Heyeping (3), Shanxi	HYP3	111.874°	38.729°	2504	32
Heyeping (4), Shanxi	HYP4	111.885°	38.730°	2435	32
Guandishan, Shanxi	GD	111.518°	37.895°	2557	32
Daqingshan, Inner Mongolia	DQ	111.259°	40.838°	2068	31
Shennongjia, Hubei	SNJ	110.258°	31.446°	2974	30
Taibaishan, Shaanxi	TB	107.806°	33.997°	3486	30
Houzhenzi, Shaanxi	HZZ	107.767°	33.950°	3350	31
Changqing, Shaanxi	CQ	107.600°	33.717°	2790	32
Helanshan (west), Ningxia	HL1	105.944°	38.838°	3428	30
Helanshan (north), Ningxia	HL2	105.944°	38.846°	2900	30
Maxianshan (west), Gansu	MXS1	103.950°	35.750°	3356	32
Maxianshan (east), Gansu	MXS2	103.950°	35.750°	3353	32
Lianhuashan, Gansu	LHS	103.750°	34.933°	2812	32
Ledu, Qinghai	LD	102.389°	36.655°	2543	32
Datong, Qinghai	DT	101.691°	36.930°	2666	32

doi:10.1371/journal.pone.0024497.t002

doi:10.1371/journal.pone.0024497.t002

DISCUSSION

Population diversity and its relationship to geographical distance and altitude

Our analysis of AFLP molecular markers indicated that R. dumulosa has maintained a high overall genetic diversity (Ht = 0.2473), similar to that of the two other species of Rhodiola but much higher than that of other perennial plants (Table 3). Thus, the expectation that genetic variability would decline due to population fragmentation was not supported in this case. This result might be observed because R. dumulosa still occurs in medium or large population sizes (40 to over 150 individuals) in some localities. Moreover, the gene flow among populations on the same mountain was high (Gst and Nm for Donglingshan and Heyeping populations). Another explanation could be that this high genetic diversity is a reflection of high historic genetic variability, which is quite common in long-lived perennial plant species [21]–[22].

Table 3. Comparison of genetic diversity of R. dumulosa with values from other Rhodiola species and selected other plant species.

Species	Genetic diversity	Marker	References conclusion
Rhodiola species			
Rhodiola dumulosa	PPL=98.22%	AFLP	Present study
	Ht=0.2473		
	I=0.3625		
Rhodiola rosea	PPL=95.3%	AFLP	high total genetic diversity [53]
	Ht=0.266		
	I=0.420		
Rhodiola fastigiata	PPL=96.61%	AFLP	high total genetic diversity [54]
	Ht=0.3329		
	I=0.4893		
Other plants			
Incarvillea younghusbandii	Ht=0.063	AFLP	low genetic diversity [55]
	I=0.096		
Cedrela odorata	PPL=98.8%	AFLP	high total genetic diversity [56]
	Ht=0.22		

doi:10.1371/journal.pone.0024497.t003

doi:10.1371/journal.pone.0024497.t003

Analyses of the correlation between population genetic diversity and geographic altitude showed a weak but significant negative correlation. The lower genetic diversity of populations at higher altitudes might be the result of the smaller populations of pollinators in these areas; R. dumulosa is thought to breed by facultative xenogamy, which requires pollinators for outbreeding [16]. However, pollination biology studies of R. dumulosa concluded that populations in open fields, such as those near peaks, received more frequent pollinator visits than those at lower altitudes [17]. Therefore, we suggest that past climatic oscillations, which may have caused the local extinction and recolonization of populations at high altitudes, could have driven this pattern. The impacts of past climatic oscillations on species ranges and genetic structure have been addressed in several previous phylogeographic studies [23]–[25]. Here, we hypothesize that the extremely low temperatures experienced during glaciation resulted in the extirpation of R. dumulosa populations in high-altitude montane regions, whereas the populations in lower altitude areas with more favorable climate conditions survived. During interglacial periods, some offspring of the surviving low altitude populations recolonized the higher altitude. Thus, populations in higher altitude sites, derived from only a few individuals from lower altitude populations, would exhibit less genetic variation. Greater genetic diversity at in situ survival areas than at recolonized areas was also found in a number of phylogeographic studies, especially in the Alps [26]–[28]. For example, the genetic variation of a widespread alpine herb, Biscutella laevigata (Brassicaceae), reflected the influence of past climate changes on the species range as a gradient of genetic diversity along recolonization pathways [27], [29]. The study of these populations also revealed that the peripheral Alps are occupied by populations with significant haplotypic variation, whereas recently glaciated areas at high altitudes in the central Alps typically contain expanding populations with a single, fixed haplotype [29]. A third potential reason for the correlation between higher altitude and lower genetic diversity pattern is simply related to population size. It is widely accepted that genetic drift can have significant effects on small populations [30]. However, we did not observe an obvious gradient of population size according to altitude, and in some sampling areas, the populations on mountain peaks are larger because the mountain has a flat top (e.g., Wutaishan).

Population structure and gene flow among naturally isolated populations

A strong correlation between genetic and geographical distances (Mantel test: $r = 0.801$; $P<0.01$) revealed a pattern of isolation-by-distance across the distribution range of R. dumulosa in China. This pattern suggested that the dispersal of this species might be constrained by distance such that gene flow is most likely to occur between neighboring populations [31]–[32]. As a result, more closely situated populations tend to be more genetically similar to one another [33]. The population genetic structure analyses (UPGMA tree, NJ tree and STRUCTURE) showed that populations tend to cluster in the same group when they are geographically restricted and occur in close proximity to one another. The congruence between the geographical distribution of populations and their genetic relationships is generally interpreted as sign of a longstanding pattern of highly restricted gene flow [34]. Our AFLP analysis revealed significant genetic differentiation and restricted gene flow among all sampled populations throughout the distribution range, which is naturally fragmented (Gst = 0.4675, Nm = 0.5695). This finding may be the result of the "ecological island" distribution pattern of R. dumulosa across its range in China. However, we also found that the gene flow among populations on the same mountain is relatively unrestricted, and the genetic differentiation among these populations is weak. For example, the Gst and Nm of the five populations in Donglingshan were 0.1125 and 3.9438, respectively, while the Gst and Nm of the four populations in Heyeping were 0.0728 and 6.6390. Furthermore, Mantel tests conducted on these two groups of populations revealed no significant correlations between genetic and geographical distances. Thus, our AFLP data suggest that although an isolation-by-distance pattern may be detected across the whole range of R. dumulosa, the gene flow and the relationship between geographical and genetic distances have different patterns at different spatial scales. Similarly distinct patterns at different spatial scales were also found for some other plant species [35].

The results of our hierarchical and model-based cluster analyses of AFLP data strongly suggested that the 35 sampled populations of R. dumulosa could be split into two clusters, one in northern China (NC) and the other in central and northwestern China (CNWC). In particular, the populations collected from northern China always clustered together regardless of the approach used for genetic structure analysis. AMOVA within the northern populations also clearly indicated that the genetic variation within populations was higher than that among populations (65.44% and 34.56%, respectively). Thus, we treated the northern populations as a reasonable cluster. However, when the populations from central and northwestern China were pooled together, AMOVA indicated that there was greater genetic variation among populations than within populations (56.90% and 43.10%), and the variation between NC and CNWC clusters was only 23.68%. Thus, the central and northwestern populations may not form a good cluster. More data may help resolve the genetic relationship between populations in these two areas. Moreover, when AMOVA was conducted based on different grouping approaches, we always found high genetic variation within populations (at least above 40%, Table 1). This finding could be due to some gene flow among populations, especially at small distribution scales, as discussed above. Alternatively, this result could reflect the preservation of ancient genetic diversity within populations as we also found significant genetic differentiation among populations or groups (Table 1). However, additional data, such as detailed gene sequence data, will be required to further dissect the evolutionary history of R. dumulosa.

In conclusion, the total genetic diversity of R. dumulosa was high, and population diversity decreased with increasing altitude. The geographical distribution of populations and their genetic relationships were consistent and most likely due to the natural geographic fragmentation of this species. A significant isolation-by-distance pattern was found across the entire distribution range in China, and significant genetic differentiation and restricted gene flow were observed among populations. However, restricted gene flow and a correlation between genetic and geographical distance were not appreciated at smaller spatial scales.

MATERIALS AND METHODS

Population sampling and DNA extraction

Thirty-five populations of R. dumulosa, including 1089 individuals (28–32 individuals per population), were sampled from 19 localities across the entire distribution of this species in 2005, 2006 and 2007. The sampling area ranged from the northernmost population

"DQ" in the Inner Mongolia Autonomous Region (40.838°N) to the southernmost population "SNJ" in Hubei province (31.446°N) and from the easternmost population "WL" in Hebei province (117.461°E) to the westernmost population "DT" in Qinghai province (101.691°E) (Fig. 5, Table 2). All sampled individuals were separated by at least five meters.

Leaf material was stored in zip-lock plastic bags with silica gel until DNA extraction. Sample vouchers were deposited in the collection at Beijing Normal University. Genomic DNA was extracted from silica gel-dried leaf material using a plant DNA extraction kit (Tiangen, Beijing, China) according to the manufacturer's protocol, with some modifications.

AFLP fingerprinting

AFLP fingerprinting was performed according to the original protocol presented by Vos et al. [36], with minor modifications. Four pairs of primers (combinations of FAM-labeled EcoRI-ATG, MseI-CTG, MseI-CAA, MseI-CAC and MseI-CAG) were used for selective amplification. Selective amplification products were separated using an ABI 3100 sequencer (Applied Biosystems) with a GeneScan ROX 500 internal size standard. Electropherograms were then analyzed using GENEMAPPER 3.7 software (Applied Biosystems). To create a binary matrix, amplified fragments of 80–500 base pairs were scored visually as having present (1), absent (0), or ambiguous (?) peaks in the output traces. Only distinct peaks were scored as present, and the manual scoring procedure was repeated twice on separate occasions to reduce scoring errors.

Genetic data analysis

The unweighted pair group method with averages (UPGMA) clustering analysis, derived from the Nei's minimum distance matrix [37], [38] as calculated in TFPGA v1.3 [39], was conducted using the SAHN module in NTSYS v2.10 [40]. One thousand bootstrapped replicate matrices of pairwise Fst among populations were calculated in AFLP-SURV. The results were used as inputs for computing Neighbor-Joining (NJ) dendrograms, using the NEIGHBOR module in the PHYLIP v3.68 [41] software package. An extended majority-rule consensus tree was produced using CONSENSE, a module for STRUCTURE v2.2 [20], adapted for dominant markers and used to assign an individual's probability of belonging to a homogeneous cluster (K populations) without prior population information. The correlated allele frequencies and admixed model were applied with a burn-in period of 100,000 and 1,000,000 MCMC replicates after burn-in. The range of clusters (K) was predefined from 1 to 35. From K = 1 to K = 10, ten runs were performed, whereas for K>10, five runs were performed. The Pr (X | K) (or "LnP(D)") can be used as an indication of the most likely number of groups, and it usually plateaus or increases slightly after the "right K" is reached [42]. Therefore, the height of the modal value of the ΔK distribution was calculated to detect the true K [42] using Structure 2(1).2-sum [18]. The similarity coefficient of pair-wise runs for each value of K was also calculated in Structure 2(1).2-sum to control the stability of the results [43]. Permutations of the most likely results among different runs for each K were conducted in CLUMPP [44]. DISTRUCT [45] was used to visualize the STRUCTURE results. The partitioning of variation at different levels was calculated by Analysis of Molecular Variance (AMOVA) in ARLEQUIN v3.01 [46] using 1,000 permutations. The 35 populations were then grouped according to the clusters indicated by the UPGMA tree, NJ dendrograms and STRUCTURE analysis results, respectively. The correlation between the genetic (Fst) and geographic population pairwise distance matrices was evaluated using a Mantel test with 9999 permutations in ARLEQUIN v3.01, and the scatter plot was constructed in SPSS v15.0. The correlation between Nei's genetic diversity and altitude was calculated using SPSS v15.0. Genetic diversity and differentiation statistics were calculated using AFLP-SURV v1.0 [47], [48]. This program estimates allele frequencies at each marker locus in each population, assuming that the markers are dominant and that there are two alleles per locus (the presence of a band is considered dominant and its absence is considered recessive). A Bayesian method with a non-uniform prior distribution of allele frequencies [49] was used to estimate the allelic frequencies, which assumed Hardy-Weinberg equilibrium. These allele frequencies were then used to analyze the genetic diversity within and between samples according to the method described by Lynch and Milligan [50]. The proportions of polymorphic loci at the 5% level (PPL), as well as the expected heterozygosity (or Nei's gene diversity, Hj), standard errors (S.E.(Hj)) and total variance (Var(Hj)) were computed for each population. Wright's fixation index, Fst [51], was computed using the Lynch and Milligan [50] method and tested with a permutation procedure of 1000 replicates. POPGENE v1.32 [52] was also used to conduct descriptive analysis for AFLP markers. Nei's gene diversity (H, analogous to Hj), the Shannon diversity index, population differentiation (Gst) and the estimate of gene flow, Nm (Nm = 0.5(1 – Gst)/Gst) were also calculated.

Table S1: Membership of each pre-defined population in each of the two clusters generated by CLUMPP based on the STRUCTURE v2.2 analysis (K=2).

Population	Sample size	Cluster (inferred gene pool)	
		1	2
WL	29	0.9960	0.0040
DL1	32	0.9970	0.0030
DL2	32	0.9960	0.0040
DL3	32	0.9900	0.0100
DL4	31	0.9530	0.0470
DL5	32	0.9770	0.0230
BH	32	0.9840	0.0160
XB1	31	0.9970	0.0030
XB2	30	0.9910	0.0090
XX	29	0.9970	0.0030
BS	32	0.9900	0.0100
WD	32	0.9180	0.0820
WZH	30	0.9980	0.0020
WX	29	0.9980	0.0020
WB1	32	0.9980	0.0020
WB2	28	0.9980	0.0020
LYS1	32	0.9580	0.0420
LYS2	30	0.9940	0.0060
HYP1	32	0.9950	0.0050
HYP2	32	0.9960	0.0040
HYP3	32	0.9960	0.0040
HYP4	32	0.9720	0.0280

GD	32	0.9150	0.0850
DQ	31	0.9080	0.0920
SNJ	30	0.2164	0.7836
TB	30	0.0458	0.9542
HZZ	31	0.0799	0.9201
CQ	32	0.0429	0.9571
HL1	30	0.1406	0.8594
HL2	30	0.1417	0.8583
MXS1	32	0.0020	0.9980
MXS2	32	0.0058	0.9942
LHS	32	0.0040	0.9960
LD	32	0.0066	0.9934
DT	32	0.0060	0.9940

Membership of each pre-defined population in each of the two clusters generated by CLUMPP based on the STRUCTURE v2.2 analysis (K = 2).

doi:10.1371/journal.pone.0024497.s001

(DOC)

Table S2□Population genetic diversity of R. dumulosa (n: population size; PPL: proportion of polymorphic loci at the 5□ level; Hj: expected heterozygosity or Nei's gene diversity; S.E.: standard error; Var: variance; H: Nei's gene diversity; I: Shannon diversity index; NC: northern China; CNWC: central and northwestern China).

Population	n	PPL	Hj	S.E.(Hj)	Var(Hj)	H	I
DT	32	27.1	0.11931	0.00947	0.000090	0.0923	0.1490
LD	32	71.6	0.18024	0.01078	0.000116	0.1440	0.2253
MXS1	32	21.3	0.09690	0.00934	0.000087	0.0742	0.1171
MXS2	32	37.3	0.13210	0.00939	0.000088	0.1044	0.1707
LHS	32	46.2	0.17898	0.01063	0.000113	0.1512	0.2399
DQ	31	77.8	0.21463	0.01157	0.000134	0.1775	0.2698
HL1	29	38.7	0.18113	0.01174	0.000138	0.1428	0.2161
HL2	29	35.6	0.16987	0.01131	0.000128	0.1320	0.2005
BS	31	39.6	0.16786	0.01098	0.000120	0.1332	0.2057

WL	29	33.8	0.15299	0.01171	0.000137	0.1259	0.1896
XB1	31	29.8	0.13086	0.01046	0.000109	0.1049	0.1643
XB2	29	33.8	0.13727	0.01023	0.000105	0.1080	0.1706
XX	28	32.0	0.12531	0.00966	0.000093	0.0993	0.1582
BH	31	40.0	0.15185	0.01046	0.000109	0.1275	0.2006
DL1	32	31.1	0.13715	0.01047	0.000110	0.1069	0.1650
DL2	31	32.9	0.15196	0.01118	0.000125	0.1205	0.1842
DL3	31	40.4	0.16991	0.01106	0.000122	0.1428	0.2218
DL4	30	69.3	0.20997	0.01164	0.000136	0.1850	0.2856
DL5	31	45.3	0.18630	0.01076	0.000116	0.1489	0.2310
HYP1	32	35.6	0.14415	0.01066	0.000114	0.1155	0.1799
HYP2	32	35.6	0.14976	0.01104	0.000122	0.1224	0.1892
HYP3	32	31.6	0.14374	0.01131	0.000128	0.1146	0.1743
HYP4	32	37.8	0.15605	0.01094	0.000120	0.1265	0.1971
LYS1	32	43.1	0.17091	0.01095	0.000120	0.1375	0.2139
LYS2	30	31.6	0.13824	0.01013	0.000103	0.1092	0.1730
GD	32	38.7	0.15521	0.01048	0.000110	0.1225	0.1919
WD	31	76.9	0.22679	0.01079	0.000117	0.2022	0.3174
WZH	29	28.9	0.12401	0.01015	0.000103	0.0982	0.1530
WX	29	31.1	0.13448	0.01048	0.000110	0.1037	0.1598
WB1	32	32.4	0.13320	0.01039	0.000108	0.1081	0.1688
WB2	28	28.4	0.12508	0.00988	0.000098	0.0969	0.1526
TB	29	25.3	0.11032	0.01002	0.000100	0.0864	0.1340
CQ	32	23.1	0.10484	0.00958	0.000092	0.0850	0.1341
HZZ	31	27.1	0.10936	0.00984	0.000097	0.0853	0.1336
SNJ	30	29.3	0.12919	0.01045	0.000109	0.1005	0.1551
Total	1089					0.2473	0.3625
NC	806					0.1972	0.3185
CNWC	283					0.2216	0.3458

Population genetic diversity of R. dumulosa. (n: population size; PPL: proportion of polymorphic loci at the 5% level; Hj: expected heterozygosity or Nei's gene diversity; S.E.: standard error; Var: variance; H: Nei's gene diversity; I: Shannon diversity index; NC: northern China; CNWC: central and northwestern China).

doi:10.1371/journal.pone.0024497.s002

ACKNOWLEDGMENTS

We thank Yong Mu, Lin Zhu, Hao-Bin Zhang and associate professor Quan-Ru Liu of Beijing Normal University for assistance with field surveys and sample collections. We are also grateful to professor Da-Yong Zhang, Yan-Ping Guo and Guang-Yuan Rao for their helpful suggestions and discussions as well as Dr. Yan-Fei Zeng, Dr. Yuan-Ye Zhang, Dr. Wan-Jin Liao and Dr. Chuan Ni for their guidance and suggestions throughout the analysis process.

Author Contributions

Conceived and designed the experiments: YH AL. Performed the experiments: YH. Analyzed the data: YH. Contributed reagents/materials/analysis tools: YH. Wrote the paper: YH.

REFERENCES

1. Young A, Boyle T, Brown T (1996) The population genetic consequences of habitat fragmentation for plants. Tree 11: 413–418. View Article PubMed/NCBI Google Scholar
2. Cardoso SRS, Provan J, Lira CDF, Pereira LDOR, Ferreira PCG, et al. (2005) High levels of genetic structuring as a result of population fragmentation in the tropical tree species Caesalpinia echinata Lam. Biodiversity and Conservation 14: 1047–1057. View Article PubMed/NCBI Google Scholar
3. Prentice HC, Lönn M, Rosquist G, Ihse M, Kindströn M (2006) Gene diversity in a fragmented population of Briza media: grassland continuity in a landscape context. Journal of Ecology 94: 87–97. View Article PubMed/NCBI Google Scholar
4. Yao XH, Ye QG, Kang M, Huang HW (2007) Microsatellite analysis reveals interpopulation differentiation and gene flow in the endangered tree Changiostyrax dolichocarpa (Styracaceae) with fragmented distribution in central China. New Phytologist 176: 472–480. View Article PubMed/NCBI Google Scholar
5. Aguilar R, Quesada M, Ashworth L, Herreriasdiego Y, Lobo J (2008) Genetic consequences of habitat fragmentation in plant populations: susceptible signals in plant traits and methodological approaches.

Molecular Ecology 17: 5177–5188. View Article PubMed/NCBI Google Scholar

6. Chen KX, Wang R, Chen XY (2008) Genetic structure of Alpinia japonica populations in naturally fragmented habitats. Acta Ecologica Sinica 28(6): 2480–2485. View Article PubMed/NCBI Google Scholar
7. Balal SR, Fore SA, Guttman SI (1994) Apparent gene flow and genetic structure of Acer saccharum subpopulations in forest fragments. Canadian Journal of Botany 72(9): 1311–1315. View Article PubMed/NCBI Google Scholar
8. Prober SM, Tompkins C, Moran CG, Bell JC (1990) The conservation genetics of Eucalyptus paliformis L. Johnson et Blaxell and E. parviflora Cambage, two rare species from south-eastern Australia. Australian Journal of Botany 38(1): 79–95. View Article PubMed/NCBI Google Scholar
9. Young AG, Merriam HG, Warwick SI (1993) The effects of forest fragmentation on genetic variation in Acer saccharum Marsh. (sugar maple) population. Heredity 71: 277–289. View Article PubMed/NCBI Google Scholar
10. Chen XY (2000) Effects of fragmentation on genetic structure of plant populations and implications for the biodiversity conservation. Acta Ecologica Sinica 20(5): 884–892. View Article PubMed/NCBI Google Scholar
11. Kuss P, Pluess AR, Ægisdóttir HH, Stöcklin J (2008) Expand+Spatial isolation and genetic differentiation in naturally fragmented plant populations of the Swiss Alps. Journal of Plant Ecology 1(3): 149–159. View Article PubMed/NCBI Google Scholar
12. Medrano M, Herrera CM (2008) Geographical structuring of genetic diversity across the whole distribution range of Narcissus longispathus, a habitat-specialist, Mediterranean narrow endemic. Annals of Botany 102(2): 183–194. View Article PubMed/NCBI Google Scholar
13. Bossuyt B (2007) Genetic rescue in an isolated metapopulation of a naturally fragmented plant species, Parnassia palustris. Conservation Biology, 21: 832–841. View Article PubMed/NCBI Google Scholar
14. Wesche K, Hensen I, Undrakh R (2006) Range-wide genetic analysis provides evidence of natural isolation among populations of the Mongolian endemic Potentilla ikonnikovii Juz. (Rosaceae). Plant Species Biology 21: 155–163. View Article PubMed/NCBI Google Scholar
15. Uhl CH (1952) Heteroploidy in Sedum rosea (L.) Scop. Evolution 6: 81–86. View Article PubMed/NCBI Google Scholar

16. Mu Y, Zhang YH, Lou AR (2007) A preliminary study on floral syndrome and breeding system of the rare plant Rhodiola dumulosa. Acta Ecologica Sinica 31(3): 528–535. View Article PubMed/NCBI Google Scholar
17. Zhu L, Lou AR (2010) Mating system and pollination biology of a high-mountain perennial plant, Rhodiola dumulosa (Crassulaceae). Journal of Plant Ecology 3(3): 219–227. View Article PubMed/NCBI Google Scholar
18. Pritchard JK, Stephens M, Donnelly P (2000) Inference of population structure using multilocus genotype data. Genetics 155: 945–959. View Article PubMed/NCBI Google Scholar
19. Falush D, Stephens M, Pritchard JK (2003) Inference of population structure using multilocus genotype data: linked loci and correlated allele frequencies. Genetics 164: 1567–1587. View Article PubMed/NCBI Google Scholar
20. Falush D, Stephens M, Pritchard JK (2007) Inference of population structure using multilocus genotype data: dominant markers and null alleles. Molecular Ecology Notes 7: 574–578. View Article PubMed/NCBI Google Scholar
21. Geert A, Rossum F, Triest L (2008) Genetic diversity in adult and seedling populations of Primula vulgaris in a fragmented agricultural landscape. Conservation Genetics 9: 845–853. View Article PubMed/NCBI Google Scholar
22. Honnay O, Jacquemyn H, Bossuyt B, Hermy M (2005) Forest fragmentation effects on patch occupancy and population viability of herbaceous plant species. New Phytologist 166: 723–736. View Article PubMed/NCBI Google Scholar
23. Beatty GE, Provan J (2011) Comparative phylogeography of two related plant species with overlapping ranges in Europe, and the potential effects of climate change on their intraspecific genetic diversity. BMC Evolutionary Biology 11: 29. View Article PubMed/NCBI Google Scholar
24. Dubreuil M, Riba M, Mayol M (2008) Genetic structure and diversity in Ramonda myconi (Gesneriaceae): effects of historical climate change on a preglacial relict species. American Journal of Botany 95: 577–587. View Article PubMed/NCBI Google Scholar
25. Beck JB, Schmuths H, Schaal BA (2008) Native range genetic variation in Arabidopsis thaliana is strongly geographically structured and reflects Pleistocene glacial dynamics. Molecular Ecology 17(3): 902–915. View Article PubMed/NCBI Google Scholar
26. Parisod C (2008) Postglacial recolonisation of plants in the western

Alps of Switzerland. Botanica Helvetica 118: 1–12. View Article PubMed/NCBI Google Scholar

27. Parisod C, Besnard G (2007) Glacial in situ survival in the Western Alps and polytopic autopolyploidy in Biscutella laevigata L. (Brassicaceae). Molecular Ecology 16: 2755–2767. View Article PubMed/NCBI Google Scholar
28. Schönswetter P, Stehlik I, Holderegger R, Tribsch A (2005) Molecular evidence for glacial refugia of mountain plants in the European Alps. Molecular Ecology 14: 3547–3555. View Article PubMed/NCBI Google Scholar
29. Parisod C, Joost S (2010) Divergent selection in trailing- versus leading-edge populations of Biscutella laevigata. Annals of Botany 105: 655–660. View Article PubMed/NCBI Google Scholar
30. Nei M, Maruyama T, Chakraborty R (1975) The bottleneck effect and genetic variability in populations. Evolution 29: 1–10. View Article PubMed/NCBI Google Scholar
31. Hutchison DW, Templeton AR (1999) Correlation of pairwise genetic and geographic distance measures: inferring the relative influences of gene flow and drift on the distribution of genetic variability. Evolution 53: 1898–1914. View Article PubMed/NCBI Google Scholar
32. Slatkin M (1993) Isolation by distance in equilibrium and non-equilibrium populations. Evolution 47: 264–279. View Article PubMed/NCBI Google Scholar
33. Wright S (1943) Isolation by distance. Genetics 28: 114–138. View Article PubMed/NCBI Google Scholar
34. Schaal BA, Hayworth DA, Olsen KM, Rauscher JT, Smith WA (1998) Phylogeographic studies in plants: problems and prospects. Molecular Ecology 7: 465–474. View Article PubMed/NCBI Google Scholar
35. Medrano M, Herrera CM (2008) Geographical structuring of genetic diversity across the whole distribution range of Narcissus longispathus, a habitat-specialist, Mediterranean narrow endemic. Annals of Botany 102: 183–194. View Article PubMed/NCBI Google Scholar
36. Vos P, Hogers R, Bleeker M, Reijans M, Lee T, et al. (1995) AFLP: a new technique for DNA fingerprinting. Nucleic Acids Res 23(21): 4407–7714. View Article PubMed/NCBI Google Scholar
37. Nei M (1972) Genetic distance between populations. Am Nat 106(949): 283–292. View Article PubMed/NCBI Google Scholar
38. Nei M (1978) Estimation of average heterozygosity and genetic distance from a small number of individuals. Genetics 89: 583–590.

View Article PubMed/NCBI Google Scholar

39. Miller MP (1997) Tools for Population Genetic Analysis (TFPGA), version 1.3. Department of Biological Sciences, Northern Arizona University, Flagstaff, AZ, USA. 16: Available: http://www.marksgeneticsoftware.net/tfpga.htm. Accessed 2011 Aug. View Article PubMed/NCBI Google Scholar
40. Rohlf FJ (2000) NTSYS-pc Numerical taxonomy and multivariate analysis system. Exeter Publishing, Ltd, Setauket, NY.
41. J (2008) PHYLIP: Phylogeny inference package, version 3.68, August 2008. Seattle, WA: University of Washington. 16: Available: http://evolution.gs.washington.edu/phylip.html. Accessed 2011 Aug. View Article PubMed/NCBI Google Scholar
42. Evanno G, Regnaut S, Goudet J (2005) Detecting the number of clusters of individuals using the software STRUCTURE: a simulation study. Mol Ecol 14: 2611–2620. View Article PubMed/NCBI Google Scholar
43. Nordborg M, Hu TT, Ishino Y, Jhaveri J, Toomajian C, et al. (2005) The pattern of polymorphism in Arabidopsis thaliana. PLoS Biol 3: 196. View Article PubMed/NCBI Google Scholar
44. Jakobsson M, Rosenberg NA (2007) CLUMPP: a cluster matching and permutation program for dealing with label switching and multimodality in analysis of population structure. Bioinformatics 23: 1801–1806. View Article PubMed/NCBI Google Scholar
45. Rosenberg NA (2004) DISTRUCT: a program for the graphical display of population structure. Mol Ecol Notes 4: 137–138. View Article PubMed/NCBI Google Scholar
46. Excoffier L, Laval G, Schneider S (2005) ARLEQUIN version 3.0: an integrated software package for population genetics data analysis. Evol Bioinform Online 1: 47–50. View Article PubMed/NCBI Google Scholar
47. Vekemans X (2002) AFLP-SURV version 1.0. Distributed by the author. Laboratoire de Génétique et Ecologie Végétale, Université Libre de Bruxelles, Belgium. 16: Available: http://www.ulb.be/sciences/lagev/aflp-surv.html. Accessed 2011 Aug. View Article PubMed/NCBI Google Scholar
48. Vekemans X, Beauwens T, Lemaire M, Roldan-Ruiz I (2002) Data from amplified fragment length polymorphism (AFLP) markers show indication of size homoplasy and of a relationship between degree of homoplasy and fragment size. Mol Ecol 11: 139–151. View Article PubMed/NCBI Google Scholar
49. Zhivotovsky LA (1999) Estimating population structure in diploids

with multilocus dominant DNA markers. Mol Ecol 8: 907–913. View Article PubMed/NCBI Google Scholar

50. Lynch M, Milligan BG (1994) Analysis of population genetic structure with RAPD markers. Mol Ecol 3: 91–99. View Article PubMed/NCBI Google Scholar
51. Hartl DL, Clark AG (1997) Principles of population genetics, 3rd edition. Sunderland, MA: Sinauer Associates.
52. Yeh FC, Yang RC, Boyle T (1997) POPGENE version 1.31, the user-friendly shareware for population genetic analysis. Molecular Biology and Biotechnology Center, University of Alberta, Edmonton, Canada. 16: Available: http://www.ualberta.ca/~fyeh/index.htm. Accessed 2011 Aug. View Article PubMed/NCBI Google Scholar
53. Meng QW (2007) Intra-species Genetic Diversity Analysis of Rhodiola Rosea L. Based on Amplified Fragment Length Polymorphism (AFLP) Marker, Xinjiang Agricultural University, China. 16: Available: http://www.lunwentianxia.com/product.sf.3204045.1/. Accessed 2011 Aug. View Article PubMed/NCBI Google Scholar
54. Lv YL (2009) Genetic Diversity Analysis of Rhodiola fastigiata in Tibet Based on Amplified Fragment Length Polymorphism (AFLP) Marker, Tibet University, China. 16: Available: http://so.med.wanfangdata.com.cn/ViewHTML/DegreePaper_D061963.aspx. Accessed 2011 Aug. View Article PubMed/NCBI Google Scholar
55. Zhu Y, Geng Y, Tersing T, Liu N, Wang Q, et al. (2009) High genetic differentiation and low genetic diversity in Incarvillea younghusbandii, an endemic plant of Qinghai-Tibetan Plateau, revealed by AFLP markers. Biochemical Systematics and Ecology 37(5): 589–596. View Article PubMed/NCBI Google Scholar
56. De la Torre A, Lopez C, Yglesias E, Cornelius JP (2008) Genetic (AFLP) diversity of nine Cedrela odorata populations in Madre de Dios, southern Peruvian Amazon. Forest Ecology and Management 255(2): 334–339.

Chapter 8

DNA DAMAGE IN PLANT HERBARIUM TISSUE

[1]Martijn Staats, [2]Argelia Cuenca, [3,4]James E. Richardson, [1]Ria Vrielink-van Ginkel, [2]Gitte Petersen, [2]Ole Seberg, [1]Freek T. Bakker

[1]Biosystematics Group, Wageningen University, Wageningen, The Netherlands

[2]Laboratory of Molecular Systematics, Natural History Museum of Denmark, University of Copenhagen, Copenhagen, Denmark

[3]Tropical Diversity Section, Royal Botanic Garden Edinburgh, Edinburgh, United Kingdom

[4]Laboratorio de Botánica y Sistemática, Universidad de Los Andes, Bogotá, Colombia

ABSTRACT

Dried plant herbarium specimens are potentially a valuable source of DNA. Efforts to obtain genetic information from this source are often hindered by an inability to obtain amplifiable DNA as herbarium DNA is typically highly degraded. DNA post-mortem damage may not only reduce the number of amplifiable template molecules, but may also lead to the generation of erroneous sequence information. A qualitative and quantitative assessment of DNA post-mortem damage is essential to determine the accuracy of molecular data from herbarium specimens. In this study we present an assessment of DNA damage as miscoding lesions in herbarium specimens using 454-sequencing of amplicons derived from plastid, mitochondrial, and nuclear DNA. In addition, we assess DNA degradation as a result of strand breaks and other types of polymerase non-bypassable damage by quantitative real-time PCR. Comparing four pairs of fresh and herbarium specimens of the same individuals we quantitatively assess post-mortem DNA damage, directly after specimen preparation, as well as after long-term herbarium storage. After specimen preparation we estimate the proportion of gene copy numbers of plastid, mitochondrial, and nuclear DNA to be 2.4–3.8% of fresh control DNA and 1.0–1.3% after long-term herbarium storage, indicating that nearly all DNA damage occurs on specimen preparation. In addition, there is no evidence of preferential degradation of organelle versus nuclear genomes. Increased levels of C→T/G→A transitions were observed in old herbarium plastid DNA, representing 21.8% of observed miscoding lesions. We interpret this type of post-mortem DNA damage-derived modification to have arisen from the hydrolytic deamination of cytosine during long-term herbarium storage. Our results suggest that reliable sequence data can be obtained from herbarium specimens.

INTRODUCTION

The world's approximately 3400 herbaria (http://sciweb.nybg.org/science2/IndexHerbariorum.asp) contain an immense number of plant specimens covering virtually all known species, making herbaria not only invaluable assets for understanding plant biodiversity [1], [2], but also largely underutilised genomic treasure troves. The expansion of next-generation sequencing (NGS) capabilities will potentially open up possibilities for cost-effective sequencing of genomes from type specimens and rare or extinct species stored in herbaria [3]. Although, DNA extraction results in irreparable damage to specimens, which conflicts with their historic and scientific importance, typically only a few milligrams of herbarium material need to be sampled. Nevertheless for small herbarium specimens (e.g. some Brassicaceae) or type specimens this can be too much, as the whole specimen basically has to be sacrificed. Therefore, considerable effort has been spent on optimizing DNA extraction protocols [4]–[6]. Furthermore, herbarium DNA is typically highly degraded into low molecular weight fragments [7]–[9]. Up until twenty years ago, herbarium specimen preparation techniques were not aimed at preserving DNA. Thus commonly used collection methods involved chemical treatments of specimens with formalin or ethanol, both of which severely affect DNA preservation in plants [7], [10], [11].

The occurrence of apuric sites, deaminated cytosine residues, and oxidized guanine residues are the main types of damage known from studies in vivo and on ancient DNA [12], [13]. In living cells, such sites can have lethal consequences and are efficiently repaired [14]. Herbarium specimen preparation, however, induces high levels of metabolic and cellular stress responses and ultimately cell death resulting in irreparably damaged DNA [15]. The post-mortem DNA damage inflicted during specimen preparation may be higher in organelles, as they are the major source of reactive oxygen species (ROS) known to inflict oxidative nucleotide damage [16], [17]. Once preserved, specimens in all major herbaria are normally (but not continuously) protected from the damaging effects of ultraviolet light and stored at moderate temperatures and at relatively low humidity, and often subjected to a two-yearly −20°C freezing cycle.

Damaged nucleotides in herbarium DNA may result in damage-specific nucleotide mis-incorporations (miscoding lesions) by DNA polymerase enzymes during amplification [18], [19]. In contrast to such polymerase-by-passable damage, strand-breaks and other DNA modifications block polymerases and thus prevent amplification. Qualitative and quantitative assessment of DNA post-mortem damage is therefore essential to determine the accuracy of DNA sequence data from herbarium specimens. The first aim of this study was to assess DNA damage as a result of polymerase non-bypassable damage using quantitative real-time PCR for multiple plastid, mitochondrial, and nuclear DNA regions. Secondly, levels of miscoding lesions in herbarium DNA were assessed using 454-sequencing of amplicons derived from each of the three genomic compartments. Using fresh and herbarium specimens of up to 114 years old, taken from the same individuals, allows a quantitative assessment of post-mortem DNA damage. Post-mortem damage was assessed, i) directly after herbarium specimen preparation by comparing results from fresh tissue and young herbarium specimens, and ii) after long-term herbarium storage by comparing results from young and old (>65 yrs.) herbarium specimens. Through statistical comparison, we investigated whether polymerase misincorporation errors alone explain the levels of miscoding lesions observed or whether they represent true damage-derived lesions in herbarium DNA. Finally, levels of DNA damage in the plastid, mitochondrial, and nuclear genomic compartments were compared in order to test for preferential DNA degradation among them.

Results

Degree of DNA fragmentation caused by herbarium preservation

The oldest herbarium material used was a 114 year old specimen of Liriodendron tulipifera. The other specimens included were Ginkgo biloba (107 years old) and Laburnum anagyroides (65 years old). The DNA extracted from these specimens was typically highly degraded and DNA fragment sizes were mostly below 1 kb (Figure S1). DNA extracts from young herbarium specimens (i.e., dated 8 July 2010) dried for 18 hours at 60°C did not contain high molecular weight DNA, and the average fragment size was below 10 kb. DNA extracts prior to filter cleanup were brownish, whereas cleanup yielded clear extracts with A260/A280 ratios between 1.7 and 2.0.

DNA yields from old and young herbarium specimens were not significantly different (P = 1.000; Table S6) and ranged between 33.91 and 103.40 ng DNA/mg tissue (Table S1). DNA yields from fresh tissue were on average 4.4–5.4 times higher, which may reflect differences in fresh versus herbarium DNA extraction efficiencies as low molecular weight DNA (DNA molecules <100 bp and nucleotides) is known to be less-efficiently recovered.

Any potential damage to herbarium DNA is likely to affect amplification efficiencies and calculated threshold cycle (Ct)-values. Depending on the kind and extent of damage this may lead to erroneous detection of gene copy numbers in herbarium DNA due to PCR jumping artefacts [28]. In this study, however, this effect was likely to be limited, because the qPCR amplicon sizes (80–140 bp) were always shorter than maximum fragments sizes (≤1 kb) in herbarium extracts. Furthermore, no amplification inhibition was observed in tests with serially diluted herbarium DNA samples (results not shown). Amplicon copy numbers for all plastid, mitochondrial, and nuclear gene regions in fresh and herbarium tissues are presented in Table S5. Mean gene copy numbers were measured by calculating the mean of plastid, mitochondial or nuclear amplicon numbers across genes per genomic compartment (Table S6).

Fresh leaf tissue yielded significantly higher plastid, mitochondrial and nuclear gene copy numbers than young and old herbarium specimens (Figure 1A–C). As measured by the fresh tissue/young herbarium gene-copy ratio, plastid copy numbers after specimen preparation were on average 9.7-fold reduced (Figure 2A), which equates to a mean reduction of 444196 plastid DNA copies (per ng of total DNA) after specimen preparation (Table S6). Compared to young herbarium specimens, plastid copy numbers were on average 3.5-fold reduced following long-term storage (young/old herbarium ratio; Figure 2B), which equates to a mean reduction of 16322 plastid DNA copies after long-term storage. Likewise, mitochondrial DNA was reduced with 31589 copies after specimen preparation and 2373 copies after long-term storage, representing an mtDNA copy number reduction of 8.6 and 3.63-fold, respectively (Figure 2).

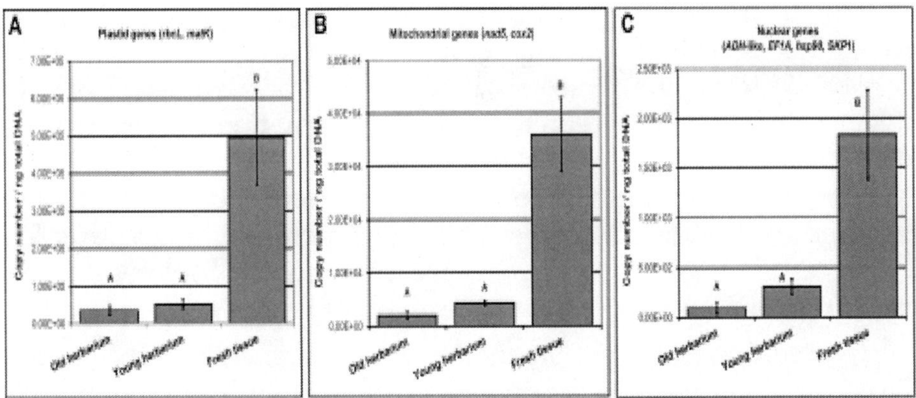

Figure 1. Copy numbers for herbarium specimens and fresh tissues for:(A) plastid genes (B) mitochondrial genes (C) nuclear genes.

Values statistically different at 5% significance level in post-hoc tests are indicated by different letters (A or B).

doi:10.1371/journal.pone.0028448.g001

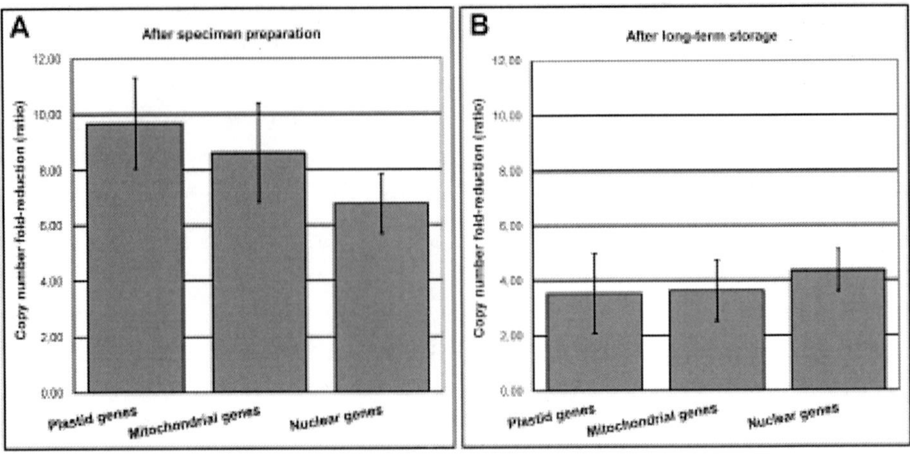

Figure 2. Copy number fold-reduction of plastid, mitochondrial and nuclear genes in herbarium specimens (A) after specimen preparation (fresh tissue/ young herbarium ratio), and (B) after long-term herbarium storage (young/ old herbarium ratio).

Values were not statistically different at 5% significance level.

doi:10.1371/journal.pone.0028448.g002

Due to the inability to produce optimal qPCR standard curves for some nuclear gene amplicons (results not shown), different sets of nuclear genes were assayed for the four species. Since the assayed loci were all predicted to be low-copy nuclear genes based on PLAZA 2.0 BLAST searches (Table S2), we believe it justifiable to compare copy numbers for different nuclear loci. Nuclear DNA was reduced with 1524 copies after specimen preparation and 215 copies after long-term storage, representing a nDNA copy number reduction of 6.8 and 4.3-fold, respectively (Figure 2).

Our results indicate that there is no difference in the degree of DNA fragmentation between plastid, mitochondrial, and nuclear genomes in herbarium specimens, as the reduction in nDNA copy numbers is not statistically different from those in plastid or mitochondrial DNA (Figure 2). Organelle DNA therefore does not appear to be preferentially degraded in herbarium tissue both directly following drying, or after long-term storage. Copy number reduction directly after drying was more pronounced than that following long-term storage. We estimate that, depending on the genomic compartment, 10.5–16.8% of the total DNA extracted from young herbarium specimens were amplifiable using Taq polymerase (Table 1). After correction for losses in DNA yield due to herbarium specimen preparation, however, we estimate the proportion of amplifiable copy numbers for plastid, mitochondrial, and nuclear DNA to be 2.4–3.8% of that of fresh controls and 1.0–1.3% after long-term storage.

Table 1. Mean DNA yield, mean gene copy numbers, and DNA yield loss-corrected copy numbers for plastid, mitochondrial and nuclear DNA relative to fresh control.

		% mean gene copy numbers[b]		
Sample type	**% mean DNA yield[a]**	**Plastid**	**Mitochondrial**	**Nuclear**
Fresh tissue	100	100	100	100
Young herbarium	22.6	10.5 (2.4)	12.2 (2.8)	16.8 (3.8)
Old herbarium	18.7	7.2 (1.3)	5.6 (1.0)	5.1 (1.0)

[a]Percentages were calculated relative to fresh controls, which were set at 100%. Mean DNA yields and mean gene copy numbers were taken from Table S6.
[b]Percentages of mean gene copy numbers corrected for DNA yield losses relative to fresh control (in brackets) are calculated as: (% mean DNA yield) × (% mean gene copy numbers).

doi:10.1371/journal.pone.0028448.t001

doi:10.1371/journal.pone.0028448.t001

Elevated levels of C T/G A transitions in old herbarium plastid DNA

The average depth of coverage per amplicon was 4640×, and therefore the level of detection was sufficient to detect 1–5% variation of single base changes (Roche application note No. 5). The overall nucleotide misincorporation error rate recorded in fresh leaf tissue was ~1.4×10−3 per nucleotide (44127 total substitutions/31933585 total nucleotides in fresh tissue; Table). This background substitution level in DNA extracted from fresh leaf tissue was assumed not only due to nucleotide misincorporations that arise from DNA polymerase errors, but also because of 454-sequencing errors and potential damage that may have arisen during DNA extraction.

Chi-square ($\chi2$) tests of independence were used to compare the observed distributions of substitutions in herbarium DNA to the expected distributions in fresh leaves. Because the amplicons were generated by PCR, the actual strand of origin of potential miscoding lesions cannot be identified. Therefore, the data were summarized into six complementary pairs of nucleotide substitution. $\chi 2$ testing of the plastid data provided strong support that the quantitative distributions of substitutions summarized over the six substitution types in young herbarium specimens ($\chi 2 = 209.37$; df = 5; $P<0.001$) and old ones ($\chi 2 = 4539.16$; df = 5; $P<0.001$) were significantly different from those in fresh tissue (Table 2). Similarly high chi-square values were observed for mitochondrial and nuclear data (Table 2).

Analyses of variance (ANOVA) were performed for each of the six substitution types separately (Table S7). This enabled the identification of specific substitution types in herbarium specimens that were over-represented and thus may be attributed to DNA post-mortem damage. C→T/G→A transitions in plastid DNA from old herbarium specimens occur significantly more frequently than in plastid DNA from fresh and young herbarium (F = 37.42; $P<0.001$; Table S7). No increased levels of C→T/G→A transitions were detected in herbarium mitochondrial and nuclear DNA. Plastid C→T/G→A transitions constitute 21.4% (4812 out of 22527 substitutions) of the observed miscoding lesions in old herbarium plastid DNA (Table 2). Although, the overall percentage of total observed nucleotide sites at which the C→T/G→A transitions occurred is low (0.0341%), the C→T/G→A rate was approximately twice as high as in fresh tissue (0.0157%; Table 2). Based on our results therefore, the estimated C→T/G→A rate in herbarium plastid DNA during storage can be calculated to be $1.53*10-6$ per nucleotide per year (Table 3). Furthermore, A→C/T→G transversions in nuclear DNA occurred significantly more frequently in old herbarium than in fresh material and young herbarium DNA (F = 10.74; P = 0.005; Table S7). However, they constitute only 3.9% (826 out of 21269 substitutions) of the total miscoding lesions, and appear to play little or no role in damage-derived miscoding lesions in herbarium DNA (Table 2).

Table 2. Estimated (C→T/G→A) rate in herbarium plastid DNA during herbarium storage.

Species	Number of substitutions per 10^6 nucleotides			
	Young herbarium	Old herbarium	Old herbarium age (yrs.)	(C→T/G→A) per nucleotide per year ± SD
Laburnum	197.55	330.02	65	$2.04*10^{-6}$
Ginkgo	229.09	384.02	107	$1.45*10^{-6}$
Liriodendron	186.14	313.52	114	$1.12*10^{-6}$
Average				$1.53*10^{-6}\pm 4.66*10^{-7}$

doi:10.1371/journal.pone.0028448.t003

doi:10.1371/journal.pone.0028448.t003

DISCUSSION

Herbaria are major depositories for conserved plant material, and their combined collections provide an unparalleled record of the world's flora [1], [29]. It is not surprising, therefore, that herbarium specimens are a common source of DNA for studies on plant evolution. Despite this, little research has focused on post-mortem DNA damage in herbarium material and its influence on data quality. As far as we know, this is the first study to quantify the prevalent damage types in herbarium DNA following an experimental approach.

Degree of DNA fragmentation caused by herbarium DNA preservation

Our results show that the most obvious form of post-mortem damage in herbarium DNA is double-stranded breaks. It is likely that the high temperatures (60–70°C) at which herbarium specimens are typically dried today causes cells to rupture quickly, concomitantly releasing nucleases, ROS, and other cellular enzymes. Such physiological conditions resemble necrosis, and this overwhelming cellular stress typically causes DNA to be degraded randomly into smaller fragments, appearing as a smear on agarose gels [30], [31]. Our experience is that DNA extracts from herbarium tissues rarely show oligo-nucleosomal DNA fragmentation [30], which is an indicator for programmed cell death (PCD). PCD is a highly coordinated death process, which is known to activate extra nucleases and ROS that damage DNA [30], [32]. Signs of DNA 'laddering' may indicate that the plant material was preserved too slowly and that the plant had experienced abiotic stresses for prolonged periods (hours, days) during preparation. Rapid desiccation of plant material has been shown to be the best way to preserve plant DNA [10], as it limits the PCD damage process.

Given that mtDNA and plastid DNA are close to major sites of ROS production and have no histones and no chromatine structure, one might expect higher levels of damage in organelle DNA during herbarium specimen preparation. However, no preferential degradation of mtDNA or plastid DNA was observed, and our results, therefore, indicate that DNA damage levels in each of the three genomic compartments are equal. The high temperatures at which herbarium specimens are typically dried most likely directly affects the degree to which DNA is preserved. Heating is known to greatly accelerate hydrolytic depurination, which in combination with the spontaneous cleavage of the phosphodiester backbone by ß-elimination will result in strand breaks [33], [34].

Our results indicate that up to 89.5% of DNA of young herbarium specimens may not be accessible to Taq polymerase (100% minus 10.5% mean plastid copy numbers; Table 1). The poor amplification success was not due to the presence of PCR inhibitors, as herbarium extracts were not found to delay amplification of an exogenous DNA control. Another possible explanation is that damage in herbarium DNA may be locus- or region-specific, and there is possibly large copy number heterogeneity in herbarium DNA samples. Gene copy numbers within each genomic compartment, however, were found to be of the same order of magnitude, rejecting a region-specific damage hypothesis. Admittedly, the representation of each genomic compartment by only two or three loci is a coarse representation of (the complexity of) organellar and nuclear genomes. A more representative sampling will be needed in order to exclude potential hotspots of degradation in herbarium DNA. A further consideration is that herbarium DNA was severely modified, not only by double stranded breaks, but also by inter-strand cross-links, abasic sites, or other structural modifications. These types of polymerase non-bypassable damage will also prevent DNA molecules from being amplified and sequenced. As an example, some studies on ancient DNA have concluded that DNA molecules were highly modified by blocking lesions [35], [36]. It will be of great value to assess the levels of abasic sites and other types of polymerase non-bypassable damage in herbarium DNA, as this will facilitate future DNA repair and rescuing strategies. Such strategies can help to improve the retrieval of accurate sequence information from herbarium specimens. For example, by pre-treating the herbarium DNA with enzymes involved in base excision repair, specific types of blocking lesions could be removed [37]. Furthermore, the use of engineered polymerases capable of extending beyond blocking lesions may allow for the amplification of highly damaged herbarium DNA molecules [38].

Degree of DNA sequence modification in herbarium DNA

Increased levels of C→T/G→A transitions were observed in plastid DNA of old herbarium specimens as compared to young ones and fresh tissues. Therefore, polymerase and sequencing errors alone cannot account for all of the observed nucleotide substitutions (as we would expect these to occur without bias), which may be better explained by added misincorporations resulting from miscoding lesions present in old herbarium DNA molecules. The observed levels were, however, low (21.4% of miscoding lesions), and we found the C→T/G→A rate to be 1.53*10−6 per nucleotide per herbarium storage year. C→T/G→A (type 2) transitions are the most common form of miscoding lesions in ancient DNA [39], [40] and are ascribed to hydrolytic deamination of cytosine to uracil [18], [41]. Our results therefore indicate that even though herbarium specimens are protected from the greatest threats to their long-term maintenance, herbarium DNA remains susceptible to hydrolytic activity. It seems probable that moisture in the air may allow for some rehydration of DNA that will lead to hydrolytic DNA damage.

We have no comprehensive explanation for the fact that the observed increased levels of C→T/G→A transitions seem to be exclusive for herbarium plastid DNA. A possible explanation is that DNA damage is probably more easily detected in plastid genes, simply because of the far higher template copy numbers compared with mitochondrial and (low-copy) nuclear genes. To what extent GC-content could play a role is not clear: %GC is usually low in angiosperm plastid genomes [42], which would seem to contradict increased type 2 transition rate hypothesis.

Further evidence for DNA damage following long-term herbarium storage is obtained from the apparent higher levels of DNA fragmentation and associated gene copy number decrease in old (65–110 yrs.) herbarium versus young herbarium specimens (Table 1; Figure S1). It should be noted, however, that the levels of DNA damage in young and old herbarium specimens may not be directly compared. First, there is no record of the exact method that had been used to prepare the old herbarium specimens. For instance, the precise temperature at which they were dried is unknown. Moreover, the temperature of commonly used gas-ovens in the early 1900 s at the Leiden herbarium would have been difficult to control. Second, we did not account for potential seasonal and year-to-year fluctuations in DNA copy numbers in old herbarium specimens (although we collected our living specimens roughly at the same original date). Plastid DNA copy numbers are known to decline during the growing season, which is typical for maturing plant leaves [43]. Therefore, a precise quantification of DNA damage after long-term herbarium storage cannot be given. However, the present results clearly indicate that nearly all DNA damage occurred during specimen desiccation, and we estimate that only a small proportion of damage can be attributed to long-term storage (Table 1; Figure 1).

Our findings may not be directly applicable to herbarium material in general, as a large variety of desiccation methods have been used to preserve specimens. Only since the 1990 s was the routine established in herbaria to collect separate DNA samples in silica gel [10], along with the herbarium specimens. Common practice for field preparation, especially in the tropics, would have been the temporary fixation with methylated spirits (the 'Schweinfurt method') or 30% formaldehyde to prevent specimens from moulding. Unfortunately, use of these chemicals is known to have destructive effects on DNA [8]. In addition, there are numerous other techniques for field drying, including the use of kerosene stoves, 100-watt light-bulbs, and air-drying on a moving vehicle. Assessing implications of each method on DNA quality will not be straightforward, as the information on the exact drying method is typically not recorded. Moreover, factors that may influence the degree of DNA preservation are likely to be species-specific (e.g., contents of secondary metabolites, etc.) and relate to the physiological state of the plant when collected, and they are therefore difficult to predict [15].

IMPLICATIONS

Our study confirms that herbaria are incredibly rich sources for reliable DNA sequence data. Various next-generation sequencing approaches can now be applied to severely fragmented DNA, and although more difficult it is still possible to generate DNA sequence data from them [44]. Nevertheless, our observation that polymerase-accessible DNA may be reduced by up to ~90% could cause challenges in sample preparation using current NGS approaches, especially when limited amounts of herbarium material are available. Probably a more serious issue would be the degree of sequence modification. Our results indicate that C→T/G→A substitutions may theoretically cause an incorrect DNA sequence to be produced; however, we predict the sequence error rate to be negligible (~0.03%) even if PCR products (i.e., the 750- bp rbcL barcode region sequence) had been sequenced from a single clone. However, the importance of these problems depends upon the type of investigation, e.g. random errors are like random noise unlikely to produce a phylogenetic signal. Indeed, so far there is no indication that, within the context of comparative studies including both fresh and old herbarium accessions, the latter share additional type 2 transitions.

The adoption of silica-gel drying of specimens generally yields higher quality DNA than most herbarium specimen preparation methods [10], and we recommend that all future collections stored for subsequent genetic analysis should continue to use this approach. However, our results confirm that herbarium DNA is a readily available resource that will be invaluable for future phylogenetic and genomic studies not least in view of the current biodiversity crisis. Other methods, e.g., proteomics or RNA studies, may require alternative conservation methods.

In a further study we will focus on elucidating the role of blocking lesions that may prevent sequencing of herbarium DNA. A study on this matter, using massive parallel sequencing of herbarium DNA, will be published elsewhere.

MATERIALS AND METHODS

Specimen sampling and herbarium specimen preparation

Plant material (leaves) of living trees were sampled from the Leiden Botanical Garden and associated herbarium specimens (65–114 years old) of the same individuals were obtained from the collections of the National Herbarium of the Netherlands at Leiden (Table S1). All necessary permissions for the described plant and specimen sampling were obtained from the respective curators, i.e. Dr. Paul J.A. Keβler (Hortus Leiden) and Dr. Jan de Koning (Herbarium Leiden). By selecting fresh and herbarium material from the same individual we avoided possible intraspecific genetic variation (e.g., genotypic variation and variations in the amounts of cpDNA) although we acknowledge that some level of somatic variation may exist. Ginkgo biloba L. (Ginkgoaceae), Laburnum anagyroides Medik. (Fabaceae), Liriodendron tulipifera L. (Magnoliaceae), and Lonicera maackii (Rupr.) Herder (Caprifoliaceae) were sampled. Fresh leaves were sampled on 8th July 2010 and stored at −80°C until DNA extraction. Herbarium specimens of fresh leaves from the same four species (i.e. dating 8 July 2010) were prepared by aluminium corrugate-drying [9] and are referred to hereafter as 'young herbarium specimens'. In order to simulate common practice in herbaria, and hence to make our results representative for typical herbarium specimens, stacked samples were pressed and dried overnight in an electric oven (Binder, type IP20) at 60°C for 18 hours. As far as we could reconstruct, our old herbarium specimens had been preserved using a similar corrugate-drying method and at similar temperature.

DNA extraction and purification

Total genomic DNA was extracted using a modified cetyl-trimethyl-ammonium-bromide (CTAB) method [20], probably the most commonly used plant DNA extraction method. Briefly, we used the following modifications: 50 mg of leaf material (with large veins removed) was weighed and placed into a 2 ml reaction tube containing five glass beads (Ø 3 mm). The reaction tube was submerged in liquid nitrogen, and the sample was then homogenized using beads (Retch, type MM2) for 30 seconds at 80 rpm. Liquid nitrogen freezing and homogenization were repeated until the leaf material had turned into a fine powder. One ml of CTAB buffer (2% CTAB, 2% PVP-40, 100 mM Tris-HCL, 1.4 M NaCl, 20 mM EDTA) and 12 µl β-mercaptoethanol were added with subsequent incubation for 60 min at 55°C. One ml of 24:1 chloroform:isoamylalcohol was then added, vortexed and centrifuged at 14.000 rpm for 4 min. The supernatant was removed, and the chloroform extraction was repeated. DNA was precipitated using 70% isopropanol at −20°C for 2 weeks, after which the DNA was pelleted at 14.000 rpm for 5 min. DNA was then re-suspended in TE buffer and treated with RNAse (Qiagen). DNA purification was performed using the Wizard DNA clean-up system (Promega Corp.) in combination with a vacuum manifold (Promega Corp.). The DNA was dissolved in 75 µl of pre-heated elution buffer (Qiagen). DNA extractions were visualized on 1% agarose gels (Figure S1), and the quantity was measured using a NanoDrop 1000 spectrophotometer (Thermo Scientific). DNA extractions were performed in duplicate or triplicate, and DNA yield per dry weight tissue was determined.

Amplification and sequencing of plastid, mitochondrial and nuclear target genes

Multiple target regions were selected for primer design, of which Expressed Sequence Tag (EST) data and/or other (partial) sequence data was available on GenBank for the plant species of interest or for closely related species. Primers were designed using Primer3Plus [21] to amplify regions of the two plastid genes (matK, rbcL) from which regions have been selected as barcode markers for land plants [22], two mitochondrial genes (coxII, nad5), and six low-copy nuclear genes (ADH-like, EF1A, H3, hsp90, RD19-like, SKP1), as well as the 18S rDNA gene (Table S2). These (high-quality and full-length) reference sequences were used for the subsequent design of nested primers for use in real-time PCR assays and 454-sequencing (see below).

PCR amplification was carried out in a 25-µl reaction mixture that contained 10 to 50 ng of total DNA isolated from fresh plant material, 1× DreamTaq™ buffer (Fermentas), 20 µg BSA, 0.2 mM of each deoxynucleoside triphosphate (Promega), 10 µM of each primer (Biolegio), and 1.0 U of DreamTaq™ DNA polymerase (Fermentas).

The following thermocycling pattern was used to amplify each gene fragment: 94°C for 5 min (1 cycle); 94°C for 30 s, 55°C for 30 s and 72°C for 60 s (35 cycles); and then 72°C for 8 min (1 cycle). PCR products were purified using the Qiaquick PCR purification kit (Qiagen) following manufacturer's instructions. In order to verify sequence identity, purified PCR products were sequenced in both directions on an ABI9600 sequencer (Greenomics).

For H3 and SKP1 of Laburnum anagyroides, ADH-like of Liriodendron tulipifera, and SKP1 of Lonicera maackii PCR products were cloned because PCR yields were insufficient for direct sequencing or because non-specific amplification products were detected. Clones were generated in E. coli using pGEM®-T Easy Vectors (Promega). Sequences were submitted to EMBL under accession numbers FR869989–FR870021. Reference gene (sub-) families, cellular component ontology, and gene family size of target regions were identified using BLAST searches against the PLAZA 2.0 platform for plant comparative genomics database at http://bioinformatics.psb.ugent.be/plaza/ [23].

Real-time quantitative PCR (qPCR)

DNA from fresh and herbarium samples was subjected to qPCR assays targeting two plastid genes (matK, rbcL), two mitochondrial genes (coxII, nad5), and three of four low-copy nuclear genes (ADH-like, EF1A, hsp90, SKP1). PCR primers for real-time PCR assays, designed to amplify fragments between 88 to 140 bp, are reported in Table S3. Real-time detection of PCR products was conducted with SYBR Green I with the MyiQ detection system (Bio-Rad) and was conducted in a total volume of 20 μl, containing 1× iQ SYBR Green Supermix (Bio-Rad), 5 μM of both primers, 20 μg BSA, and 5 to 10 ng of total DNA. The temperature profile was: 95°C for 5 min (1 cycle); 95°C for 15 s, 60°C for 20 s, and 72°C for 20 s (40 cycles), with a final extension step at 72°C for 5 min. After amplification, melting curve analysis was performed from 60°C to 95°C, with increments of 0.5°C per 10 seconds. Real-time PCR standards were prepared by cloning target amplicons and subsequent amplification with M13 forward and reverse primers. The cloned PCR fragments were then purified and the yield was quantified using a NanoDrop 1000 (Thermo Scientific). Samples were analysed in triplicate (replicates of same DNA), and serial-diluted samples and no-template controls were included. Gene copy numbers were quantified absolutely by quantifying them per nanogram of total DNA. Amplification efficiencies for optimized qPCR assays calculated from slopes of the standard curves are shown in Table S3: efficiencies were between 1.90 and 2.09, with R2-values, a measure for the quality of the regression fit, between 0.92 and 0.99. An exogenous control sequence from the wingless gene (Wg) of Cymothoe Hüber 1819 sp. (Lepidoptera, Insecta) was spiked to test for inhibition of qPCR assays by substances in the plant DNA extracts, but no inhibition was observed (data not shown). Gene copy numbers were considered to be duplicate (e.g., rbcL and matK, or nad5 and cox2) or triplicate measurements (ADH-like, EF1A, and hsp90 in Liriodendron tulipifera) in one-way ANOVA analyses, performed in PASW Statistics version 18.0.2 (SPSS Inc.). Copy number fold-reductions of plastid, mitochondrial, and nuclear genes directly after specimen preparation were expressed as 'fresh tissue/young herbarium gene copy number ratio', whereas those after long-term storage were expressed as 'young herbarium/old herbarium gene copy number ratio'.

454-sequencing of amplicons

Fusion primers for unidirectional 454 sequencing of amplicons of two plastid genes (matK, rbcL), two mitochondrial genes (coxII, nad5), six low-copy nuclear genes (ADH-like, EF1A, H3, hsp90, RD19-like, SKP1) and the 18S rDNA gene were constructed that incorporated the GS FLX Titanium forward primer A and a 10-bp multiplex identifier (MID) or reverse primer B (Table S4). The forward primer (Primer A-key) was: 5'-ccatctcatccctgcgtgtctccgactcag-MID-template-specific-sequence-3'; and the reverse primer (Primer B-key) was 5'-cctatcccctgtgtgccttggcagtctcag-template-specific-sequence-3'. MIDs 1 to 4 from the standard 454 set (Roche Technical Bulletin No. 013-2009) were selected, and each of the four species used was marked using a specific MID.

PCR amplification was conducted in a total volume of 50 µl, containing 1× High Fidelity PCR Buffer (Invitrogen), 5 mM of each deoxynucleoside triphosphate (Promega), 10 µM of each primer (Biolegio), 50 mM MgSO4, 1.0 U of Platinum®Taq High Fidelity, 20 µg BSA and 5 to 10 ng of total DNA using the following thermo profile: 94°C for 1 min (1 cycle); 94°C for 30 s, 55°C for 30 s and 68°C for 60 s (35 cycles); and then 68°C for 8 min (1 cycle).

Amplicons were purified using the Agencourt AMPure XP kit (Beckman Coultier) and quantified using the Quant-iT PicoGreen dsDNA Assay Kit (Invitrogen). Products from the same treatment (i.e. 'fresh tissue', 'young herbarium', or 'old herbarium') were mixed at equimolar concentrations to ca. 1*107 molecules/μl. For each amplicon pool emPCR was performed following the emPCR Method Manual Lib-L for medium volume (Roche, 2009) and sequenced in a quarter of a PicoTiterPlate™ using a GS FLX Titanium Series. Pyrosequencing was carried out with primer A using the Genome Sequencer FLX (Roche Life Sciences Technology) at the Natural History Museum, University of Copenhagen, Denmark.

Amplicon sequence analysis was performed using the Galaxy platform [24], [25] and included the following: first, a plot reflecting the base quality distribution was constructed. For each treatment, raw 454 sequence data and quality scores were converted to FASTQ format [26]. Low quality 3-prime ends were trimmed using a fixed value of 5% of the sequence length. Clean reads were reconverted to FASTA format and filtered by MID identifiers, constructing one file for each species-treatment.

Subsequenctly, reads were mapped against a reference sequence under the Roche-454 98% identity criterion and only reporting matches above 90% of identity and covering at least 50% of the reference sequence. A BAM file was generated from the SAM output, to only report bases with at least 100× coverage. The filtered pile up was examined for regions with a drastic drop in coverage, such as at the end of a homopolymer or at the end of the sequence, and these regions were manually removed from the output.

In sequences of EF1A, hsp90, H3, and SKP1 a minority of nucleotide positions were found to be polymorphic in fresh controls and were omitted from further analysis. We assumed these to result from either allelic variation or to be paralogous sequence variants (not shown).

Nucleotide substitutions were counted across reads and grouped into six substitution types that are effectively indistinguishable in PCR reactions [27]; (A→C/T→G), (A→G/T→C), (A→T/T→A), (C→A/G→T), (C→G/G→C) and (C→T/G→A). The numbers of observed substitutions among the herbarium data was scaled to match the total amount of nucleotides in data from fresh tissue. For example, the corrected young herbarium count of plastid data for substitution pair (A→G/T→C) was calculated as: (Observed young herbarium plastid A→G/T→C)*(Total nucleotides in data fresh tissue plastid DNA)/(Total nucleotides in data young herbarium plastid DNA) = (11634*10746437)/14193327 = 8808.65. Chi-square ($\chi 2$) tests of independence (at $p \geq 0.05$ level) were used to investigate whether the distributions of substitutions over the six substitution types were the same in sequence data obtained from fresh and herbarium DNA. The test was made two-sided because a priori the substitution direction from fresh DNA is not known. One-way analyses of variance (ANOVA) were performed to identify which of the six nucleotide substitution types occur at highest rates in each genomic compartment in herbarium DNA. For ANOVA, substitutions counts were expressed as numbers of substitutions per million total nucleotides observed (across reads).

ACKNOWLEDGMENTS

The authors would like to thank Jan de Koning for his help in finding suitable herbarium specimens. Gerben Bijl is acknowledged for technical assistance.

Author Contributions

Conceived and designed the experiments: MS FTB. Performed the experiments: MS AC RVG. Analyzed the data: MS AC. Contributed reagents/materials/analysis tools: MS AC GP OS. Wrote the paper: MS AC JER FTB.

REFERENCES

1. Bebber DP, Carine MA, Wood JRI, Wortley AH, Harris DJ, et al. (2010) Herbaria are a major frontier for species discovery. Proc Natl Acad Sci USA 107: 22169–22171.
2. Sebastian P, Schaefer H, Telford IRH, Renner SS (2010) Cucumber (Cucumis sativus) and melon (C. melo) have numerous wild relatives

in Asia and Australia, and the sister species of melon is from Australia. Proc Natl Acad Sci USA 107: 14269–14273.

3. Metzker ML (2010) Sequencing technologies – the next generation. Nat Rev Genet 11: 31–46. View Article PubMed/NCBI Google Scholar
4. Cubero OF, Crespo A, Fatehi J, Bridge PD (1998) DNA extraction and PCR amplification method suitable for fresh, herbarium-stored, lichenized, and other fungi. Pl Sys Evol 216: 243–249. View Article PubMed/NCBI Google Scholar
5. Drábková L, Kirschner J, Vlcek C (2002) Comparison of seven DNA extraction and amplification protocols in historic herbarium specimens of Juncaceae. Pl Mol Biol Rep 20: 161–175. View Article PubMed/NCBI Google Scholar
6. Telle S, Thines M (2008) Amplification of cox2 (~620 bp) from 2 mg of up to 129 years old herbarium specimens, comparing 19 extraction methods and 15 polymerases. Plos One 3: e3584. View Article PubMed/NCBI Google Scholar
7. Doyle JJ, Dickson EE (1987) Preservation of plant species for DNA restriction endonuclease analysis. Taxon 36: 715–722. View Article PubMed/NCBI Google Scholar
8. Pyle MM, Adams RP (1989) In situ preservation of DNA in plant specimens. Taxon 38: 576–581. View Article PubMed/NCBI Google Scholar
9. Harris SA (1993) DNA analysis of tropical plant species: an assessment of different drying methods. Pl Syst Evol 188: 57–64. View Article PubMed/NCBI Google Scholar
10. Chase MW, Hills HH (1991) Silica gel: an ideal material for field preservation of leaf samples for DNA studies. Taxon 40: 215–220. View Article PubMed/NCBI Google Scholar
11. Srinivansan M, Sedmak D, Jewell S (2002) Effect of fixatives and tissue processing on the content and integrity of nucleic acids. Am J Pathol 161: 1961–1971. View Article PubMed/NCBI Google Scholar
12. Lindahl T (1993) Instability and decay of the primary structure of DNA. Nature 362: 709–715. View Article PubMed/NCBI Google Scholar
13. Pääbo S, Poinar H, Serre D, Jaenicke-Despres V, Hebler J, et al. (2004) Genetic analyses from ancient DNA. Annu Rev Genet 38: 645–679. View Article PubMed/NCBI Google Scholar
14. Lindahl T, Wood RD (1999) Quality control by DNA repair. Science 286: 1897–1905. View Article PubMed/NCBI Google Scholar
15. Savolainen V, Cuénoud P, Spichiger R, Martinez MDP, Crèvecoeur

M, et al. (1995) The use of hebarium specimens in DNA phylogenetics: evaluation and improvement. Pl Syst Evol 197: 87–98. View Article PubMed/NCBI Google Scholar

16. Roldán-Arjona T, Ariza RR (2009) Repair and tolerance of oxidative DNA damage in plants. Mut Res 681: 169–179. View Article PubMed/NCBI Google Scholar
17. Boesch P, Weber-Lotfi F, Ibrahim N, Tarasenko V, Cosset A, et al. (2011) DNA repair in organelles: pathways, organization, regulation, relevance in disease and aging. Biochim Bioph Acta 1813: 186–200. View Article PubMed/NCBI Google Scholar
18. Gilbert MTP, Hansen AJ, Willerslev E, Rudbeck L, Barnes I, et al. (2003) Distribution patterns of postmortem damage in human mitochondrial DNA. Am J Hum Genet 72: 48–61. View Article PubMed/NCBI Google Scholar
19. Stiller M, Green RE, Ronan M, Simons JF, Du L, et al. (2006) Patterns of nucleotide misincorporations during enzymatic amplification and direct large-scale sequencing of ancient DNA. Proc Natl Acad Sci USA 103: 13578–13584. View Article PubMed/NCBI Google Scholar
20. Doyle JJ, Doyle JL (1987) A rapid DNA isolation procedure for small quantities of fresh leaf tissue. Phyt Bull 19: 11–15. View Article PubMed/NCBI Google Scholar
21. Untergasser A, Nijveen H, Rao X, Bisseling T, Geurts R, et al. (2007) Primer3Plus, an enhanced web interface to Primer3. Nucleic Acids Res 35: W71–W74. View Article PubMed/NCBI Google Scholar
22. Hollingsworth PM, Forrest LL, Spouge JL, Hajibabaei M, Ratnasingham S, et al. (2009) A DNA barcode for land plants. Proc Natl Acad Sci USA 106: 12794–12797. View Article PubMed/NCBI Google Scholar
23. Proost S, Van Bel M, Sterck L, Billiau K, Van Parys T, et al. (2009) PLAZA: a comparative genomics resource to study gene and genome evolution in plants. Pl Cell 21: 3718–3731. View Article PubMed/NCBI Google Scholar
24. Blankenberg D, Von Kuster G, Coraor N, Ananda G, Lazarus R, et al. (2010) Galaxy: a web-based genome analysis tool for experimentalists. Curr Protoc Mol Biol 19: 10.1–10.21. View Article PubMed/NCBI Google Scholar
25. Goecks J, Nekrutenko A, Taylor J, The Galaxy Team (2010) Galaxy: a comprehensive approach for supporting accessible, reproducible, and transparent computational research in the life sciences. Genome Biol 25: R86. View Article PubMed/NCBI Google Scholar
26. Blankenberg D, Gordon A, Von Kuster G, Coraor N, Taylor J, et al.

(2010) Manipulation of FASTQ data with Galaxy. Bioinform 26: 1783–1785. View Article PubMed/NCBI Google Scholar

27. Hansen AJ, Willerslev E, Wiuf C, Mourier T, Arctander P (2001) Statistical evidence for miscoding lesions in ancient DNA templates. Mol Biol Evol 18: 262–265. View Article PubMed/NCBI Google Scholar
28. Pääbo S, Irwin DM, Wilson AC (1990) DNA damage promotes jumping between templates during enzymatic amplification. J Biol Chem 265: 4718–4721. View Article PubMed/NCBI Google Scholar
29. Jobba LN, Roberts DL, Pimm SL (2011) How many species of flowering plants are there? Proc Biol Sci 278: 554–559. View Article PubMed/NCBI Google Scholar
30. Reape TJ, Molony EM, McCabe PF (2008) Programmed cell death in plants: distinguishing between different modes. J Exp Bot 59: 435–444. View Article PubMed/NCBI Google Scholar
31. McCabe PF, Levine A, Meijer PJ, Tapon NA, Pennell RI (1997) A programmed cell death pathway activated in carrot cells cultured at low cell density. Plant J 12: 267–280. View Article PubMed/NCBI Google Scholar
32. Gill SS, Tuteja N (2010) Reactive oxygen species and antioxidant machinery in abiotic stress tolerance in crop plants. Plant Physiol Biochem 48: 909–930. View Article PubMed/NCBI Google Scholar
33. Lindahl T, Nyberg B (1972) Rate of depurination of native deoxyribonucleic acid. Biochem 11: 3610–3618. View Article PubMed/NCBI Google Scholar
34. Lindahl T (1990) Repair of intrinsic DNA lesions. Mutation Res 238: 305–311. View Article PubMed/NCBI Google Scholar
35. Hansen AJ, Mitchell DL, Wiuf C, Paniker L, Brand TB, et al. (2006) Crosslinks rather than strand breaks determine access to ancient DNA sequences from frozen sediments. Genetics 173: 1175–1179. View Article PubMed/NCBI Google Scholar
36. Heyn P, Stenzel U, Briggs AW, Kircher M, Hofreiter M, et al. (2010) Road blocks on paleogenomes - polymerase extension profiling reveals the frequency of blocking lesions in ancient DNA. Nucleic Acids Res 38: e161. View Article PubMed/NCBI Google Scholar
37. Briggs AW, Stenzel U, Meyer M, Krausse J, Kircher M, et al. (2010) Removal of deaminated cytosines and detection of in vivo methylation in ancient DNA. Nucleic Acids Res 38: e87. View Article PubMed/NCBI Google Scholar
38. Shapiro B (2008) Engineered polymerases amplify the potential of

ancient DNA. Trends Biotechnol 26: 285–287. View Article PubMed/NCBI Google Scholar

39. Brotherton P, Endicott P, Sanchez JJ, Beaumont M, Barnett R, et al. (2007) Novel high-resolution characterization of ancient DNA reveals C > U-type base modification events as the sole cause of post mortem miscoding lesions. Nucleic Acids Res 35: 5717–5728. View Article PubMed/NCBI Google Scholar
40. Briggs AW, Stenzel U, Johnson PLF, Green RE, Kelso J, et al. (2007) Patterns of damage in genomic DNA sequences from a Neandertal. Proc Natl Acad Sci USA 104: 14616–14621. View Article PubMed/NCBI Google Scholar
41. Hofreiter M, Jaenicke V, Serre D, von Haeseler A, Pääbo S (2001) DNA sequences from multiple amplifications reveal artifacts induced by cytosine deamination in ancient DNA. Nucleic Acids Res 29: 4793–4799. View Article PubMed/NCBI Google Scholar
42. Smith DR (2009) Unparalleled GC content in the plastid DNA of Selaginella. Plant Mol Biol 71: 627–639. View Article PubMed/NCBI Google Scholar
43. Rowan BA, Oldenburg DJ, Bendich AJ (2009) A multiple-method approach reveals a declining amount of chloroplast DNA during development in Arabidopsis. BCM Plant Biol 9: 3. View Article PubMed/NCBI Google Scholar
44. Millar CD, Huynen L, Subramanian S, Mohandesan E, Lambert DM (2008) New developments in ancient genomics. Trends Ecol Evol 7: 386–393.

Citations

CHAPTER 1

Strigens A, Schipprack W, Reif JC, Melchinger AE (2013) Unlocking the Genetic Diversity of Maize Landraces with Doubled Haploids Opens New Avenues for Breeding. PLoS ONE 8(2): e57234. doi:10.1371/journal.pone.0057234

CHAPTER 2

Van Zonneveld M, Scheldeman X, Escribano P, Viruel MA, Van Damme P, et al. (2012) Mapping Genetic Diversity of Cherimoya (Annona cherimola Mill.): Application of Spatial Analysis for Conservation and Use of Plant Genetic Resources. PLoS ONE 7(1): e29845. doi:10.1371/journal.pone.0029845

CHAPTER 3

Sun X, Ma P, Mumm RH (2012) Nonparametric Method for Genomics-Based Prediction of Performance of Quantitative Traits Involving Epistasis in Plant Breeding. PLoS ONE 7(11): e50604. doi:10.1371/journal.pone.0050604

CHAPTER 4

El-Kassaby YA, Cappa EP, Liewlaksaneeyanawin C, Klápště J, Lstibůrek M (2011) Breeding without Breeding: Is a Complete Pedigree Necessary for Efficient Breeding? PLoS ONE 6(10): e25737. doi:10.1371/journal.pone.0025737

CHAPTER 5

Chen Y, Mao Y, Liu H, Yu F, Li S, et al. (2014) Transcriptome Analysis of Differentially Expressed Genes Relevant to Variegation in Peach Flowers. PLoS ONE 9(3): e90842. doi:10.1371/journal.pone.0090842

CHAPTER 6

Sim S-C, Van Deynze A, Stoffel K, Douches DS, Zarka D, et al. (2012) High-Density SNP Genotyping of Tomato (Solanum lycopersicum L.) Reveals Patterns of Genetic Variation Due to Breeding. PLoS ONE 7(9): e45520. doi:10.1371/journal.pone.0045520

CHAPTER 7

Hou Y, Lou A (2011) Population Genetic Diversity and Structure of a Naturally Isolated Plant Species, Rhodiola dumulosa (Crassulaceae). PLoS ONE 6(9): e24497. doi:10.1371/journal.pone.0024497

CHAPTER 8

Staats M, Cuenca A, Richardson JE, Vrielink-van Ginkel R, Petersen G, et al. (2011) DNA Damage in Plant Herbarium Tissue. PLoS ONE 6(12): e28448. doi:10.1371/journal.pone.0028448

INDEX